Distillation Symposium 2020

Topical Conference at the 2020 AIChE Spring Meeting and 16th Global Congress on Process Safety

Online
17 – 21 August 2020

ISBN: 978-1-7138-2324-7

Printed from e-media with permission by:

Curran Associates, Inc.
57 Morehouse Lane
Red Hook, NY 12571

Some format issues inherent in the e-media version may also appear in this print version.

Printed with permission by Curran Associates, Inc. (2021)

For permission requests, please contact AIChE
at the address below.

AIChE
120 Wall Street, FL 23
New York, NY 10005-4020

Phone: (800) 242-4363
Fax: (203) 775-5177

www.aiche.org

Additional copies of this publication are available from:

Curran Associates, Inc.
57 Morehouse Lane
Red Hook, NY 12571 USA
Phone: 845-758-0400
Fax: 845-758-2633
Email: curran@proceedings.com
Web: www.proceedings.com

Distillation Symposium 2020

Topical Conference at the 2020 AIChE Spring Meeting and 16th Global Congress on Process Safety

Online
17 – 21 August 2020

AIChE
The Global Home of Chemical Engineers

TABLE OF CONTENTS

591161 2020 AIChE Spring Meeting & 16th GCPS AGILE Award Keynote Address - Beyond the Horizon: The Factors and Forces Affecting the Global Petrochemical Industry

Abstract:

Over the past 10 years, LyondellBasell and the global petrochemical industry has undergone a dramatic transformation. In this era of unprecedented disruption, it is likely that the next the decade will present both challenges and opportunities for the petrochemical sector. LyondellBasell CEO Bob Patel will summarize the company's progress to date and provide a forward-looking perspective on key opportunities and risks faced by the industry in 2020 and beyond.

591157 Securing Enterprises via Cybersecurity Understanding, Resiliency, and Education

Abstract:

Securing Enterprises via Cybersecurity Understanding, Resiliency, and Education

Industry 4.0 initiatives are enabled by leveraging the deployment of information technologies into the operational technology environment. While a logical approach in defining the future of smart manufacturing, it remains fret with challenges and risks that must be mitigated assure enterprise resilience to cyber and cyber-physical attacks. As economies recalibrate in the aftermath of the current public health emergency, definitions of what constitutes "critical infrastructure" are broadening while, at the same time, attack surfaces and the threat landscape are expanding. This means companies must place a greater emphasis on taking the steps and making the investments necessary to overcome the challenges and navigate these challenges and risks to maximize their resilience and secure and protect their enterprises. This presentation will explore the link between cybersecurity and process safety in the context of a safety culture. It will also identify cybersecurity challenges in industry 4.0 initiatives, the integration of information technologies (IT) and operational technologies (IT), the delivery of training, and the need for credentialing and ongoing professional development.

AIChE Paper 83c

TROUBLESHOOTING ONCE-THROUGH REBOILERS

Clifford Alan Maat
Senior Staff Engineer
TPC Group

Sasha Sudar
Senior Engineer
TPC Group

Kevin Bellard
Senior Production Coordinator
TPC Group

Jonathan Barton
Senior Production Coordinator
TPC Group

Joe Martin
Staff Engineer
TPC Group

Prepared for Presentation at the 2020 Spring National Meeting
Distillation Symposium 2020: Chemicals and Petrochemicals Distillation and Operations I
Houston, Texas (Virtual), August 17, 2020

Abstract: Once-through reboilers provide greater thermal and separation efficiency than conventional circulating thermosiphon reboilers; however, they are subject to different modes of failure. Examples of operating problems with once-through reboilers are reviewed and recommendations made for prevention and mitigation.

Background

Once-through reboilers are often preferred to conventional circulating thermosiphon reboilers where process conditions allow. The reasons are

- **Greater Thermal Efficiency** – Because once-through reboilers receive liquid from the bottom tray, which is cooler than the tower bottoms, the temperature difference from the heating medium is greater, and more heat can be transferred with the same surface area.
- **Greater Separation Efficiency** – For some tower bottoms compositions, especially wide-boiling mixtures, a once-through reboiler significantly improves the quality of separation by partially vaporizing the liquid from the bottom tray before any of it contacts the tower bottoms [1].

However, once-through reboilers can fail in ways that circulating thermosiphon reboilers do not. Generally, these failures either cause bottom tray liquid to bypass the reboiler [2] or back liquid up onto the bottom tray.

Column A – Creative Solution Failed

Column A processes a wide-boiling mixture using a once-through reboiler.

During one campaign, the reboiler performance deteriorated significantly, and fouling was proven to not be the problem. It was not possible to mitigate the poor performance online.

Upon shutdown, the panels of the bottom tray were found to be fitting poorly, allowing substantial leakage to the tower bottom, bypassing the reboiler.

Figure 1 – Leaking Bottom Tray

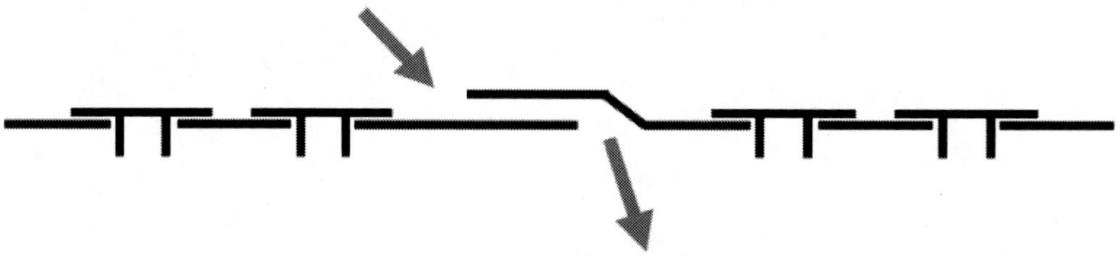

To minimize the leakage, gaskets were installed:

Figure 2 – Gaskets Installed In Bottom Tray

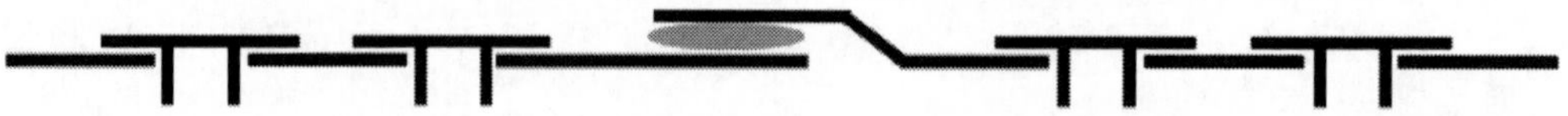

During the next campaign, the reboiler deteriorated to even worse performance. Upon shutdown, it was discovered that the gasket material had interacted with the process chemicals to create an aggressive polymer which bent the tray panels and caused even greater leakage:

Figure 3 – Polymer Growth at Gaskets In Bottom Tray

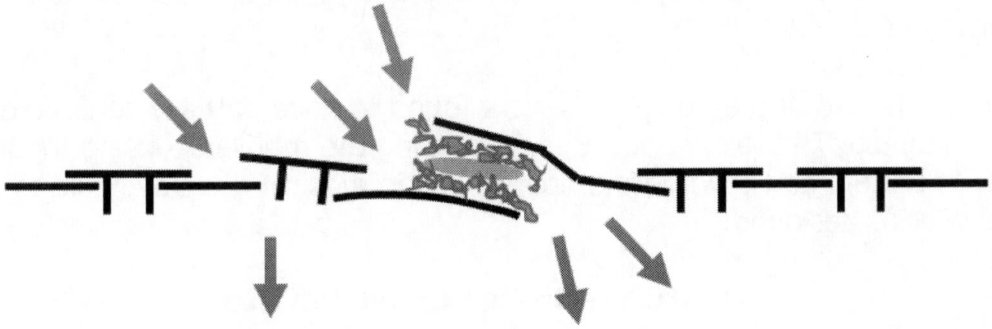

The gaskets were removed and the tray panels carefully repaired.

Column B – Raising Tower Level Enables Continued Operation

Column B separates a low-volatility solvent from C4 hydrocarbon using four pass floating valve trays and a once-through reboiler with three shells in parallel. The bottom tray collects all the liquid and directs it into two troughs. The two troughs direct the liquid to a single collection box at one side of the column. The box extends downward to the level of the outlet nozzles feeding the three reboilers.

Figure 4 – Normal Operation Column B

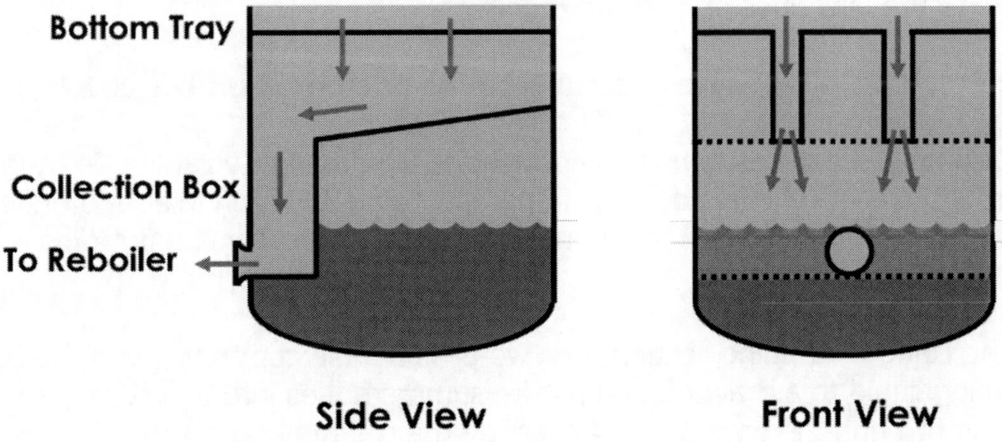

During one campaign, the reboiler performance deteriorated significantly:

- Insufficient duty
- Low steam flow
- Steam control valve wide open
- High reboiler outlet temperature

The tube bundles were removed and found to be clean. The level transmitters were confirmed to be functioning. Raising the tower bottom level to over 90% improved reboiler performance sufficiently to continue operation until the next scheduled shutdown.

Upon shutdown, two of the box panels were found to have partially separated, allowing liquid to fall directly to the tower bottom, bypassing the reboiler. Raising the liquid level in the tower bottom had covered the hole, enabling most of the tray liquid to continue to the reboiler as designed.

Figure 5 – Broken Collection Box

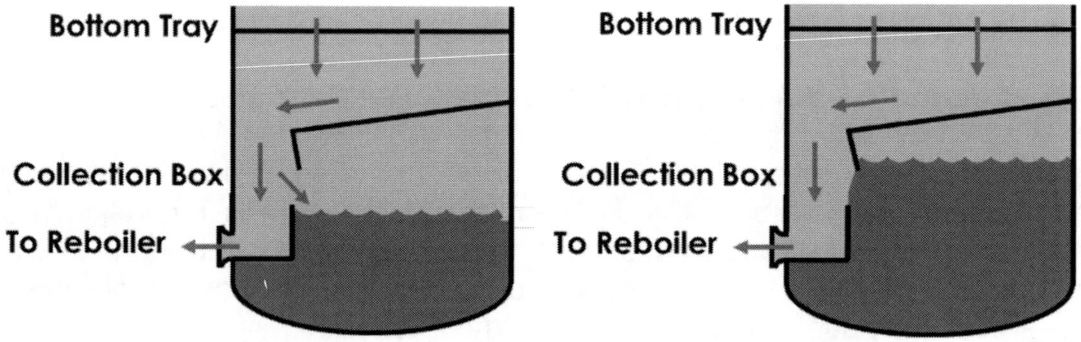

The box panels were repaired and reinforced before startup. Also, piping was installed to enable tower bottoms liquid to be circulated from the bottoms pump discharge to the inlet lines of all three reboiler shells.

Column C – No Startup Line or Bottoms Pump - 2017

Column C separates a multicomponent oil from a high-volatility C4 product. Two once-through steam reboilers operate in parallel using common inlet and outlet pipes. Live steam is also injected to assist in removal of C4 and lower the bottoms temperatures.

The column stripping section has two pass bubble cap trays, with the bottom tray draining liquid to a draw-off sump. The sump empties into a vertical pipe exiting the column through the bottom head. Outside the column and a short distance away, the pipe divides into two and then feeds two horizontal steam reboilers via bottom shellside nozzles. Live steam is injected into either or both reboiler inlet pipes. The reboiler return pipes begin at the top shellside nozzles, then join together and return to the tower shell underneath the bottom tray. The tower bottoms flows through a level control valve and a heat recovery exchanger into a surge drum at low pressure. From the surge drum, the liquid is pumped through coolers and back to the process.

Figure 6 – Normal Operation Column C

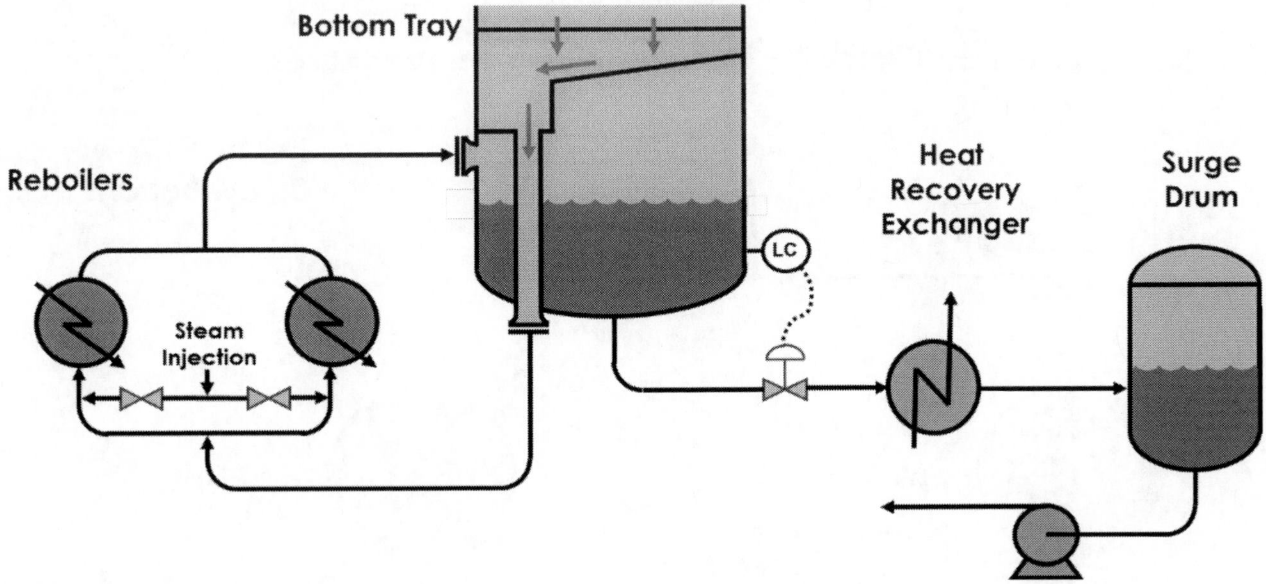

Early in one campaign, the column was unable to properly remove the C4 from the bottoms product:

- Substantial product losses
- Reduction in unit capacity
- Bottoms temperature 10-20 degrees below normal
- Reboiler outlet temperature hotter than normal
- Reboiler duty 10-30% below normal
- Steam control valve wide open
- Steam side pressure in the reboiler shell almost as high as the steam header pressure

Hole in the Internal Pipe to the Tower Nozzle Feeding the Reboiler? – Referring to Column B above, if the problem was due to a hole in the pipe, and the hole was above the normal liquid level, a higher liquid level might cover the hole. Raising the liquid level in the tower bottom did not improve reboiler performance.

Broken Bottom Tray? – A tower scan showed normal froth levels on the stripping section trays, though froth height was lower on one side of the bottom tray.

Dirty Reboilers? – One of the reboiler tube bundles was removed and inspected; it showed no unusual fouling. There was no significant loss of performance by removing one reboiler bundle; therefore, heat transfer surface area was proven to not be the problem.

Unusual temperature pattern: The live steam is controlled with an automatic flow control valve and then normally fed to both reboilers via manual gate valves. When one reboiler was taken out of service, the manual gate valve to the out-of-service shell was closed. It was left closed for some time after the reboiler tube bundle was reinstalled. The following pipe surface temperature pattern was observed:

Figure 7 – Reboiler Piping Temperatures

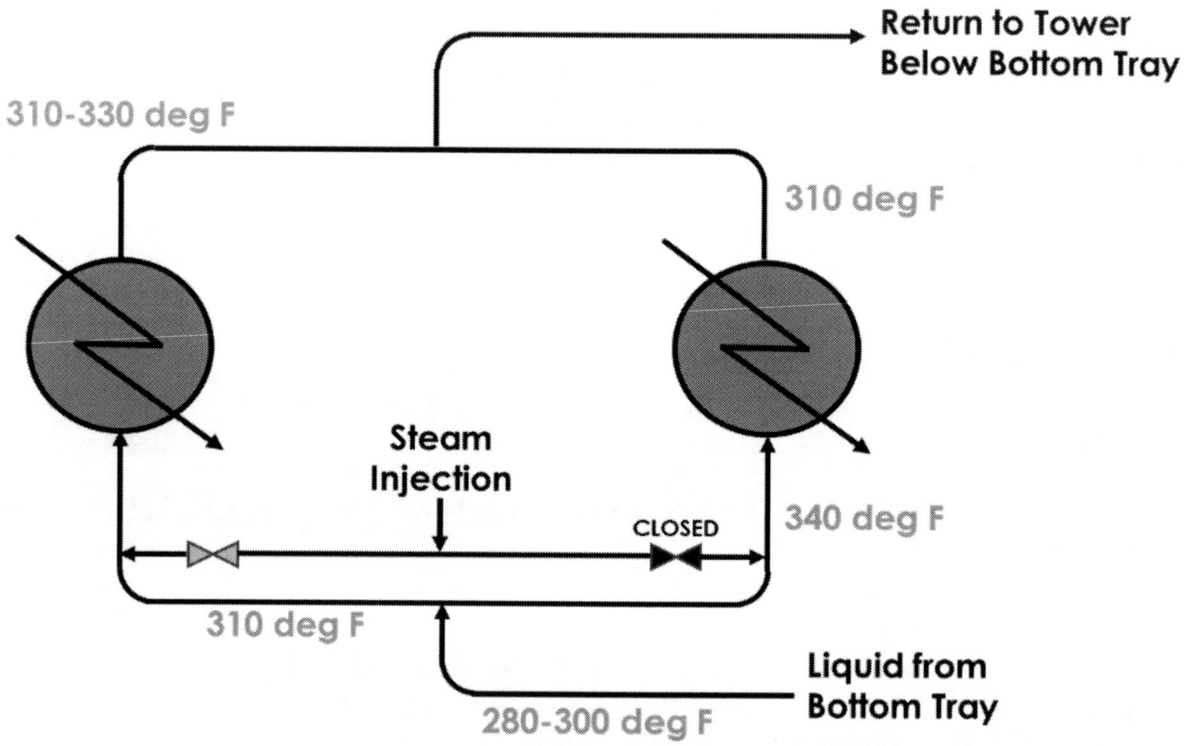

The injected steam was inducing flow upward through the east reboiler, but the west reboiler was flowing backwards. Reopening the steam valve to the west reboiler restored the proper flow direction.

Restriction in the reboiler inlet pipe? Partial plugging in the reboiler inlet pipe would account for low flow to the reboiler, but it would also raise the liquid level in the box to the bottom tray. Properly functioning bubble cap trays do not allow drainage of liquid to the tray below (except for a small amount through drain holes), so the downcomer to the bottom tray, and then all the downcomers in the tower, would be expected to flood. No tower flooding was observed; the column pressure drop was below normal. Also, a piping scan showed low level in the inlet pipe.

Attempted Mitigation by Partial Forced Circulation – Concluding that liquid from the bottom tray was somehow flowing directly to the tower bottoms (bypassing the reboiler), bottoms flow was forced into the reboiler inlet using a hose from the discharge of the downstream oil pump:

Figure 8 – Partial Forced Circulation

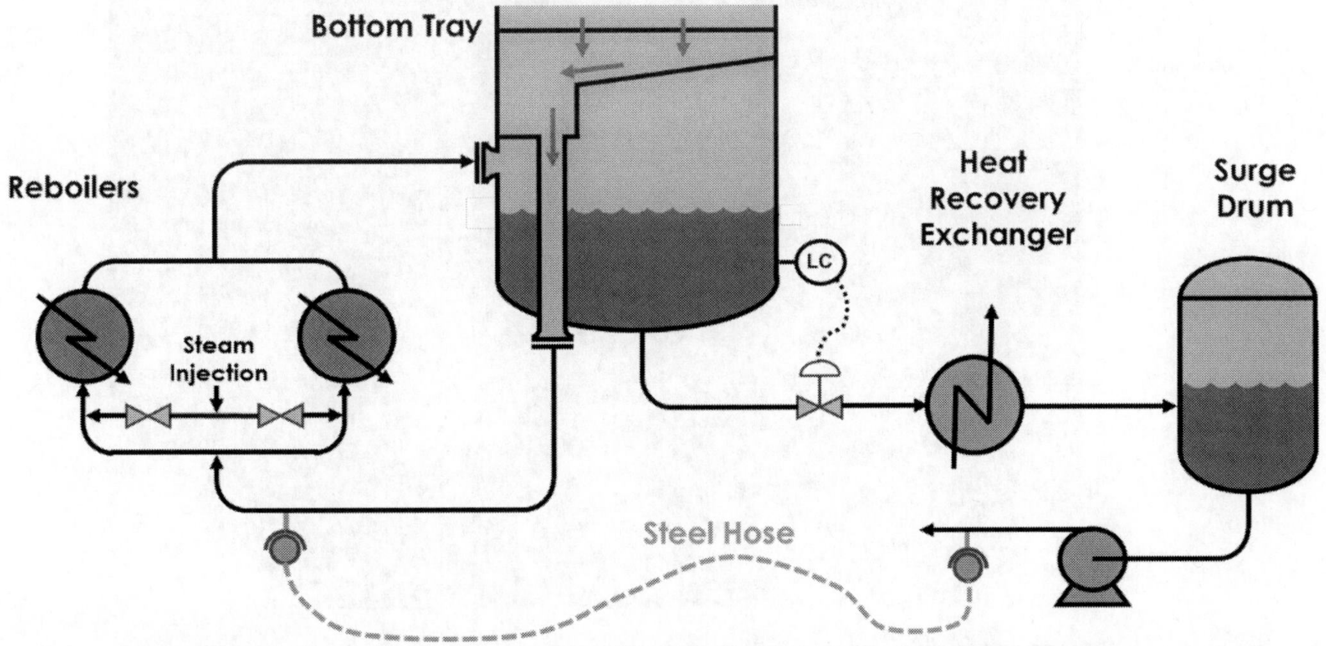

Steam flow to the reboiler increased, but tower performance did not substantially improve. Much of the reboiler duty was being used to reheat the subcooled liquid feed, so there was only a small increase in boilup. Process side effects included loss of control at the level control valve (100% open) and undesirable temperature increase at the oil surge drum.

Operating Adjustments – Suboptimal but adequate performance was restored by increasing the column feed preheater duty, reflux, and live steam flow.

Discovery at Shutdown – Upon unit shutdown and entering the tower, tray panels were discovered dislodged and sitting in the tower bottom:

Figure 9 – Damaged Bottom Tray

Column D – Bottoms Pump Available for Partial Forced Circulation - 2018

Column D separates a multicomponent oil from a high-volatility C4 product using bubble cap trays. The liquid from the bottom tray is directed to a collection box and then through an internal pipe exiting the side of the tower to a once-through steam reboiler. Live steam is also injected to the reboiler inlet pipe to assist in removal of C4 and lower the bottoms temperatures.

Figure 10 – Normal Operation Column D

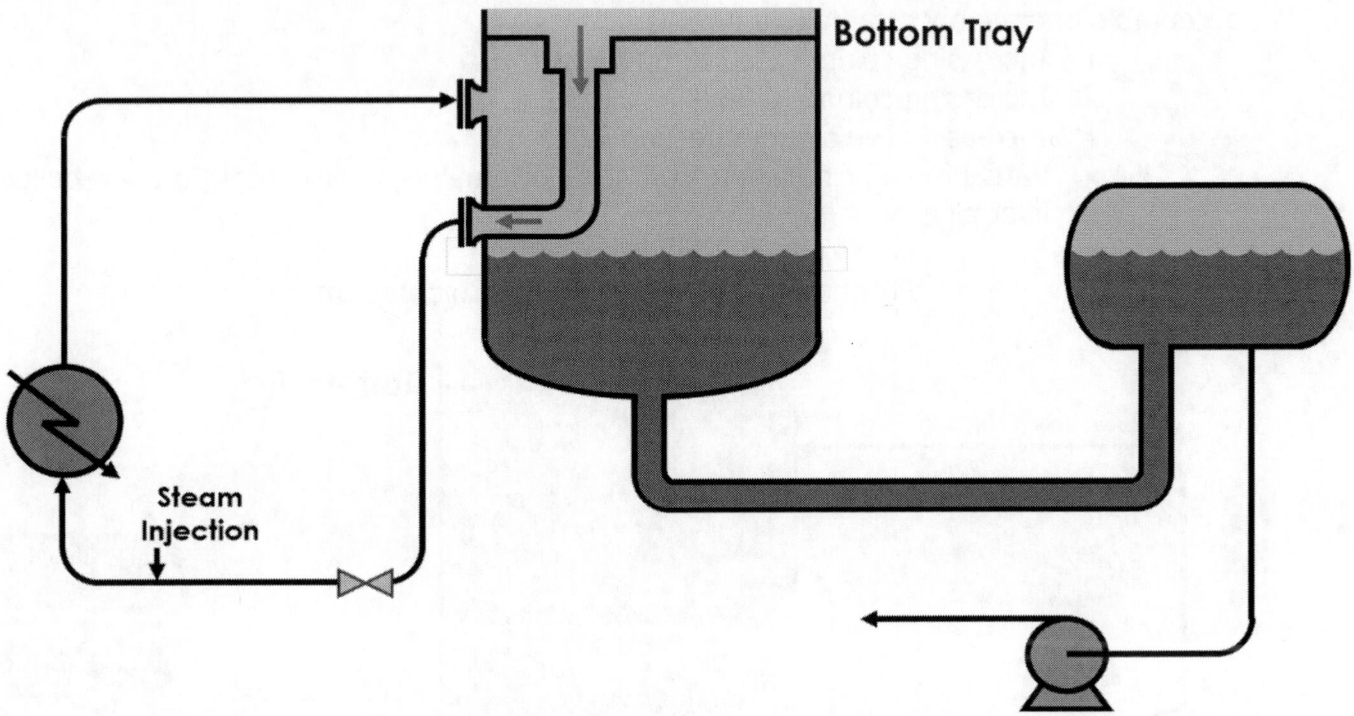

The reboiler outlet piping returns to the tower bottom, the vapor going up to the bottom bubble cap tray and the liquid exiting through the tower bottoms line to a surge tank. From the surge tank, bottoms pumps send the liquid back to the process.

Discovery of Tray Damage – During a routine shutdown, the bottom tray was found to be cracked and could not be repaired. The only trays available for purchase on short notice were sieve or fixed valve trays. A fixed valve tray was installed.

Inability to Start Up – Upon startup, the reboiler was unable to provide the required duty:

- Reboiler outlet temperature hotter than normal.
- Tower bottoms temperature lower than normal
- Steam valve wide open

- Steam side pressure in the reboiler shell almost as high as the steam header pressure

Tray Dumping at Low Vapor Rate? Insufficient vapor flow to a fixed valve tray can result in liquid dumping through the tray. This could, in turn, starve the reboiler of liquid and cause low boilup and low vapor flow [3] [4]. The live steam flow to the reboiler was temporarily increased to ensure that the tray liquid would be directed to the collection box, but the reboiler still did not work properly.

Second Replacement of Bottom Tray – The column was shut down again, the fixed valve tray removed, and the existing third-from-the-bottom tray moved down to the bottom tray location. The reboiler showed similar suboptimal symptoms, but acceptable operation was obtained after

- Increasing column feed temperature
- Increasing column reflux
- Increasing live steam injection
- Partial forced circulation from the bottoms pump discharge to the reboiler inlet pipe

Figure 11 – Partial Forced Circulation

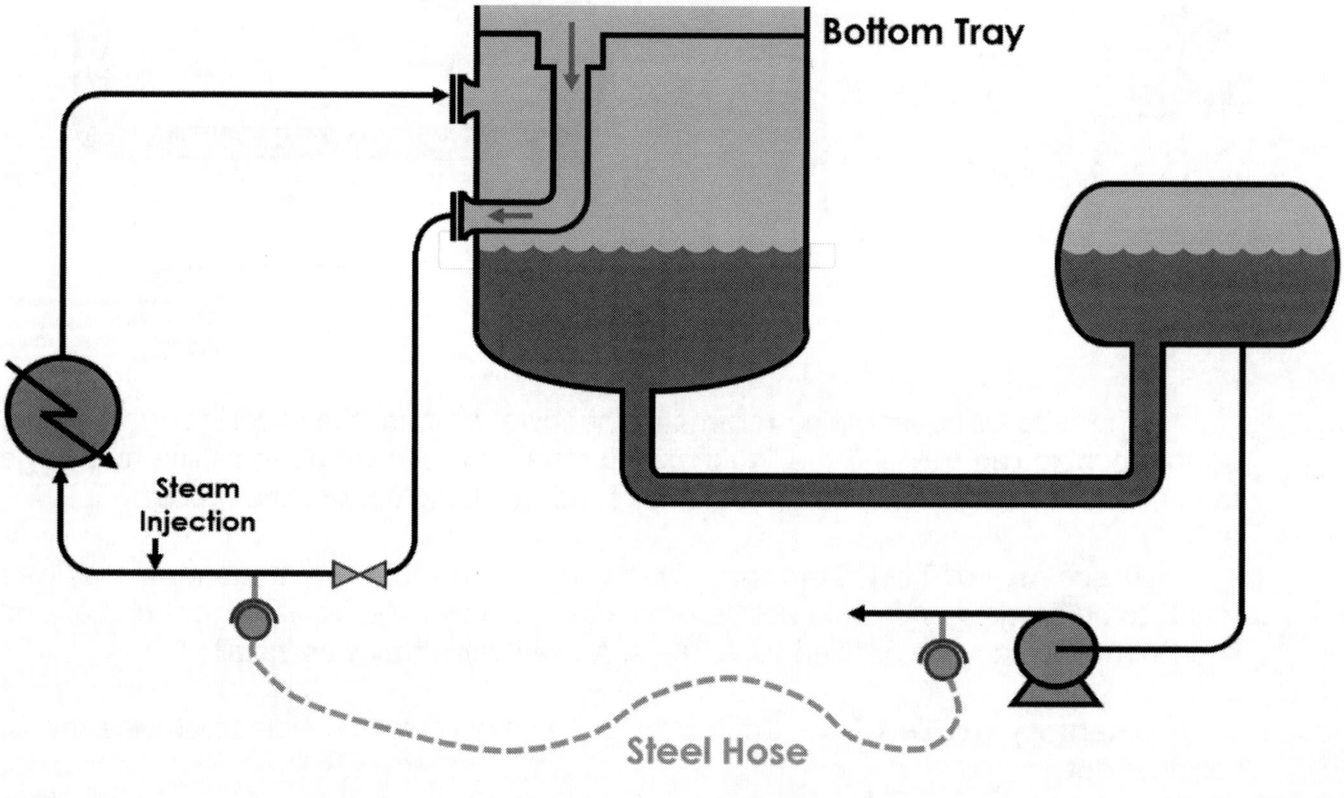

Tower Scan Ambiguous – Tower scanning indicated froth on the bottom tray, possible level in the tray/box, and low level in the piping between the tower and the reboiler inlet.

Leakage Confirmed through Tray/Box/Pipe to Tower Bottom – One troubleshooting test was to close a manual gate valve in the pipe between the tower and the reboiler inlet, while maintaining the forced circulation from the bottoms pump through the reboiler. Some liquid backed up in the box/pipe, but the trays did not flood, and a new, similar steady state was achieved with the valve closed. It can be inferred that there must be a large leak into the bottom of the tower, bypassing the reboiler.

Column C – 2019 - Same Problem Again

Reboiler performance is again worse than usual. Tower scans show that the bottom tray is holding much less liquid than the normal trays above:

Figure 12 – Scan of Column C

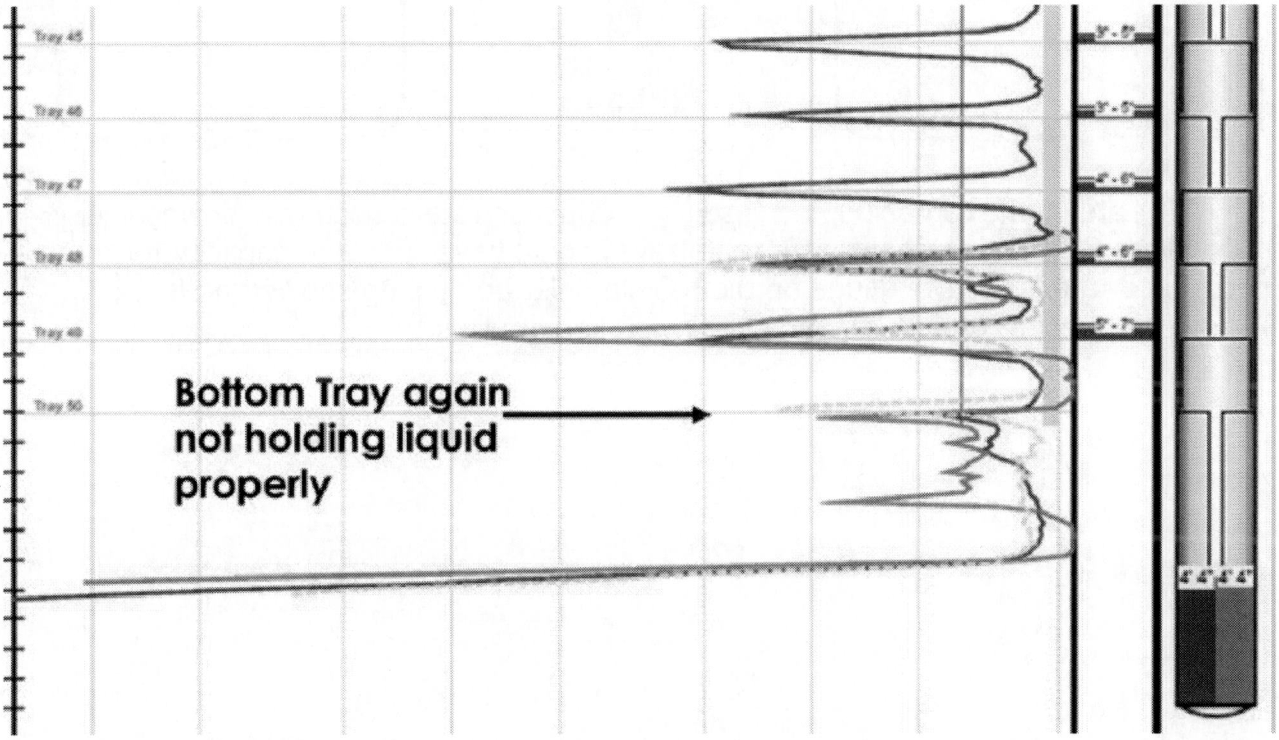

Scans also showed very little liquid in the pipe to the reboiler inlets.

Upon shutdown, tray panels from the bottom tray were again found on the bottom of the tower.

Column E –Trays Flooding

Column E separates a low-volatility solvent from C4 hydrocarbon using two pass floating valve trays and a once-through reboiler with two shells in parallel. Normally there is low or no level in the local gauge on the reboiler inlet box/piping.

Figure 13 – Normal Operation Column E

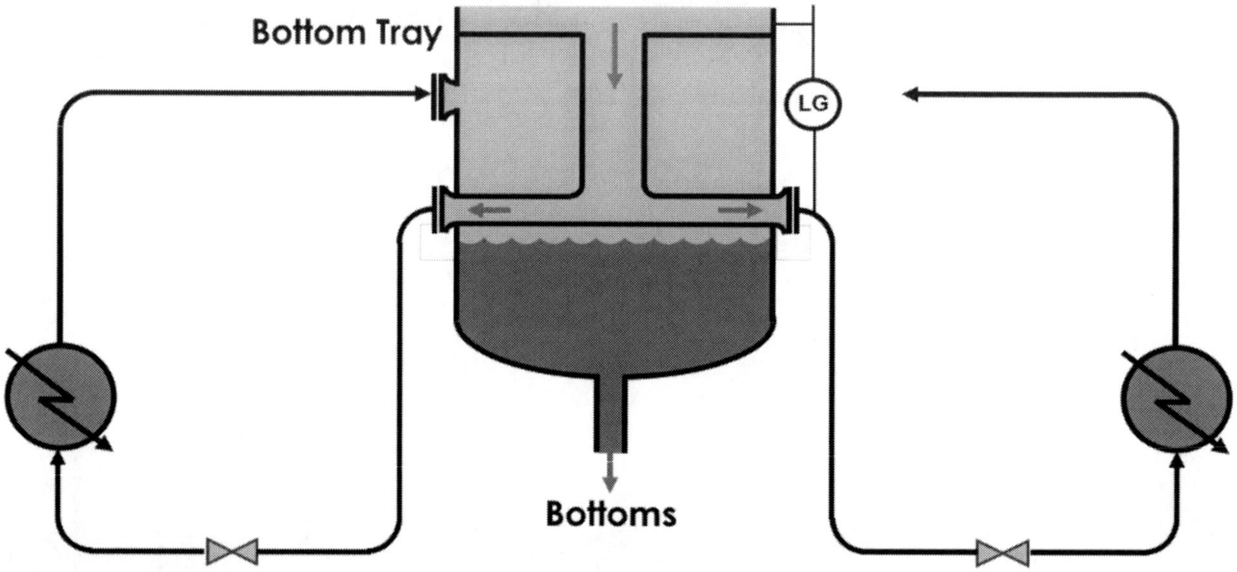

Late in one campaign, increases in production rate caused tray downcomer flooding. Calculations showed that the trays should have sufficient capacity for the required rate. The level gauge on the reboiler inlet box/piping showed no level; however, upon thorough cleaning it showed liquid full.

Figure 14 – Liquid Backed Up to Trays

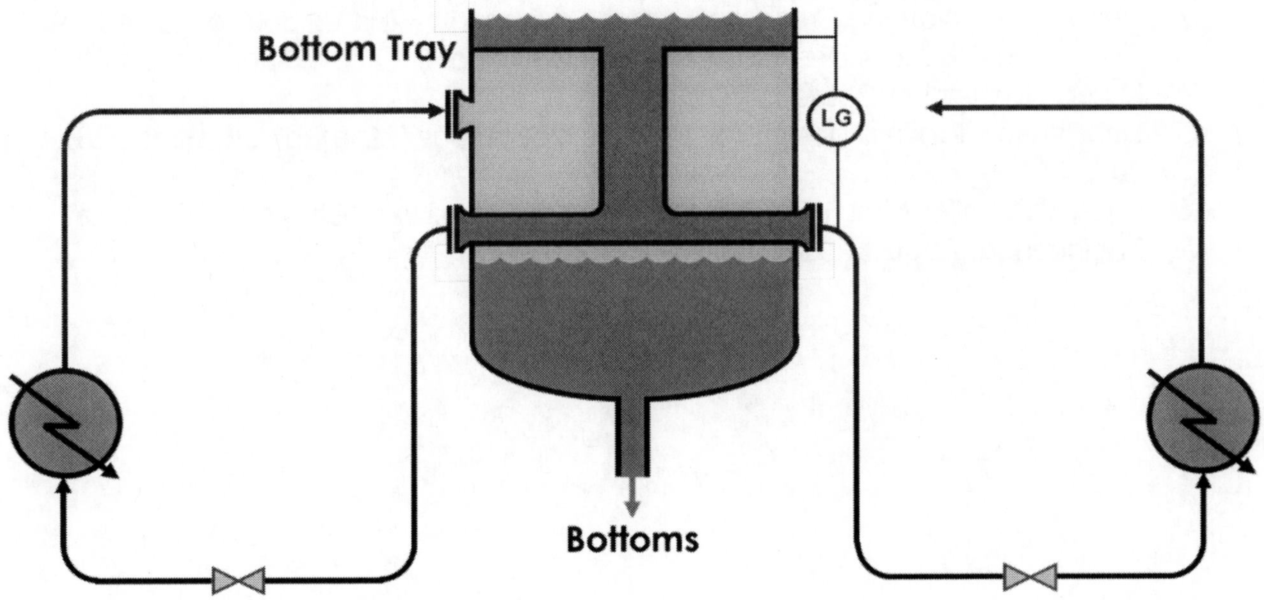

Fouling in the reboiler shells was causing high pressure drop and backing liquid up to the bottom tray. The tube bundles were removed and cleaned, and normal operation was restored.

Operating and Design Recommendations for Once-Through Reboilers

1. Remove free water from the column before heating up with hydrocarbons/oil [5]
2. Use strong materials for collection boxes and routing pipe
3. Make the bottom tray a strong chimney tray rather than an active tray
4. Enable online conversion to a circulating reboiler by installing either
    - A startup line from the tower bottom to the reboiler inlet, creating a circulating thermosiphon [6], or
    - A recirculation line from a hot bottoms pump to the reboiler inlet, creating a forced circulation

References

1. Lieberman, "Working Guide to Process Equipment," 2nd edition, pages 463-468.
2. Lieberman, "Working Guide to Process Equipment," 2nd edition, pages 48-49.
3. Lieberman, "Process Design for Reliable Operations," 2nd Edition, pages 89-90.
4. Kister, "Distillation Operation," page 101.
5. Lieberman, "Process Design for Reliable Operations," 2nd Edition, pages 201-204.
6. Kister, "Reboiler Circuit Debottleneck with No Hardware Changes," Chemical Engineering, August 2017.

590357 Reuse and Relocation: A Case History

Abstract:

Potentially, used equipment can save time and money over new equipment in your plant. Occasionally, used equipment can be picked up for very low prices. But getting it to the plant and properly working does require skills that differ from purchasing new equipment.

Buying used instead of new may be less expensive and quicker. Experience has shown that often the cost savings have been minor or in some cases costs even more than new equipment. In contrast, the schedule savings in procurement are often been significant.

First, this discussion will distinguish between used equipment and remanufactured equipment. Remanufactured equipment is material taken out of a plant, then taken off to a shop and reconditioned and returned to a 'like-new' state. The reconditioning may be either by the original equipment manufacturer (OEM) or a specialist vendor. Most specialist vendors focus one type of equipment. Many remanufacturing vendors focus one specific equipment types. Common equipment available as remanufactured or reconditioned includes control valves and pumps. The specialist vendors typically use after-market replacement components instead of OEM parts. The specialist vendor takes the risk of the reconditioning cost and usually provides some type of warranty with the equipment.

In contrast, used equipment is generally purchased with minimum cleaning and reconditioning. No warranty is supplied by the seller. Various purchase methods and situations apply. The equipment may be purchases as-is, where is, or may be purchased in 'after-cleaning'. This depends upon the agreement between the buyer and seller.

A large amount of used equipment on the market comes from shut-down or bankrupt facilities. Also, it may have sat unused for years, or even decades. Equipment purchased this way rarely comes with a warranty or performance guarantee of any type. Additionally, it often lacks design documentation. Both mechanical and process design information may be missing. At best, some mechanical design information is available. As long as the name plate is identifiable and the original manufacturer is still in business, it may be possible to get original design mechanical details from them. However, this request will often require payment. Additionally, the effect of repairs or modifications since the equipment left the original manufacturer may remain mysterious.

Finally, the process requirements for the new service are nearly always different from the original process requirements. Expert-level knowledge may be required to decide if the equipment will really do the required job.

Adding repair, reconditioning, and engineering costs to the used equipment cost is what makes saving money with it so difficult. If you're fortunate enough to find equipment with a very close match to what you need, then you can save money. In contrast, significant time can be saved. Currently some specialty items can have 48-week or longer deliveries from order. If you can find the right equipment, it might be at the plant in a few weeks instead. Even if you have to pay more, getting the plant running again is worth a lot.

This paper examine a plant that purchased a 42-inch diameter tower 'used' from a pharmaceutical plant. It had been built and erected but never placed into service. The unit the tower was in was abandoned part-way through construction. Some years latter it was purchased with the intent that it was to be used as a distillation tower in a solvent recovery unit.

In spite of the original owner still being in business, the tower was sold with no documentation on the internals inside it. No mechanical or process details were available on internals configuration or original design rates, conditions, or physical properties. The tower was purchased, loaded onto a truck (horizontally) and shipped well over a thousand miles to its new home. It was unloaded and erected in the new plant. At no time was the internal condition of the tower inspected.

After the new plant started up, the tower failed to perform. Troubleshooting quickly ran into problems. What were the internals? What was the mechanical range of liquid and vapor rates the internals could handle? What was the mechanical condition of the internals? No one knew.

The paper examines the history of the tower, troubleshooting, and the problems that had to be solved.

589637 Extractive Distillation Tower Troubleshooting after Internals Revamp

Abstract:

Extractive Distillation Tower Troubleshooting after Internals Revamp

Bailee Yarbrough, Senior Process Engineer, Process Engineering Separations Member, Dow Inc., Freeport, TX

Chet Davidson, Process Engineering Separations Expertise Area Leader, Dow Inc., Freeport, TX

Tim Cooper, Global Technology Leader, Dow Inc., Plaquemine, TX

E-mail Addresses:

BYarbrough@dow.com

DDavidson@dow.com

WTCooper1@dow.com

Extractive distillation can make difficult separations feasible with the addition of a solvent, making it an efficient separation method for close boiling components. After a revamp at a Dow plant, an extractive distillation tower was unable to meet the design rates and desired product recovery. The tower had been reconfigured from consisting of all packed internals to including a section of high capacity trays. Immediate flooding after start-up resulted in a shutdown and reinspection of the internals, which showed no damage or fouling. After the second start-up, erratic column behavior continued, temperatures swinging as much as 30 degrees Celsius in one section of the tower, in a matter of minutes. The tower proved difficult to control and still exhibited flooding tendencies at higher rates, even though these rates were almost 15% below the design rate. Improper instrumentation installation led to misleading pressure drop measurements.

Our team was tasked with troubleshooting the tower to provide stable operation and determine how to get the tower towards design rates and target recovery of the product. An assortment of troubleshooting techniques was used to aid in diagnosing the tower including gamma scanning and various lab experiments, which ultimately led to the discovery of the root cause. The gamma scan showed both maldistribution and a rapidly changing density during the scan timeframe, consistent with the temperature fluctuation readings. However, the scan showed no evidence of flooding or severe fouling at the time of the scans. This presentation will focus on the tower's problems, troubleshooting techniques including the interpretation of the gamma scans, and ultimately the root cause of the premature flooding. It will also highlight the impact of changing internals from packing to trays.

589430 Perils of Careless Handling of Phase Equilibria in Chemical Systems

Abstract:

This paper discusses the all-too-common pitfalls, in process simulation work for complex chemical mixtures, when using "built-in" binary interaction parameters for activity coefficient models in commercial process simulators. Part of the problem arises from the fact that the data sources used by the vendors are either undocumented or extremely out of date and do not include the latest information. In some cases, these built-in activity coefficient interaction parameters are based on group contribution methods that could easily be 30-100% in error. Such situations can lead to disastrously wrong results for VLE, LLE, or VLLE systems, especially when the phase equilibria for the most important binaries in the mixture have been characterized improperly.

In this paper, we will discuss the proper methods and procedures to ensure reliable results. We will also discuss guidelines for requesting fresh phase equilibria measurements in the laboratory when available data are insufficient or inaccurate. A few industrial examples will be presented to illustrate these ideas and procedures.

590340 Wash Oil Control:Inferred Versus Measured Rates

Abstract:

Crude oil distillation towers use a wash section to remove entrainment from the feed. This section requires wash (wetting) liquid to maximize performance and prevent the wash section from coking. This is a difficult control problem. The liquid rate has a sharp optimum. Too low a rate, and the wash section cokes, shutting down the unit. Too high a rate and significant yield or capacity losses can result. Being wrong in either direction has a high cost.

The liquid rate can be controlled by either direct measurement and control or by inferring a rate based on energy and mass balance around the unit and controlling a secondary stream. The direct measurement system has higher capital and operating costs. The indirect method has lower capital and operating costs, but has a higher risk of upsets and larger uncertainties.

This presentation reviews the results of Monte Carlo modeling of a crude unit to evaluate the economic benefits of the direct approach versus the uncertainties in the indirect approach.

591159 When Disasters Happen: A Collaborative Approach Towards the Community's Recovery Process

Abstract:

The 2020 Virtual AIChE Spring Meeting & 16th GCPS Wednesday keynote "When Disasters Happen: A Collaborative Approach Towards the Community's Recovery Process" will be presented by Thomas Munoz, Emergency Manager and Homeland Security Director with Texas City. As the Emergency Manager and Homeland Security Director with Texas City, the home of 19 chemical plants, he is responsible for the safety of all its citizens along with managing the crisis communication plan with local industry and major stakeholders.

Safety and disaster planning is vital to the recovery process after a major incident. As key critical infrastructure industries, issues may arise that have not been planned for in advance, such as the recovery from an incident. Developing and maintaining partnerships with local partners in your community will support addressing unforeseen challenges. Recovery efforts include planning with key partners and understanding the phases of recovery, which also consider the mental health and well of everyone involved in the response, as well as the community impacted. In this keynote address a discussion of pre-incident planning considerations and strategies as well as lessons learned/after action improvements post-recovery will be discussed. Additionally, collaboration considerations regarding local resources to promote recovery in the event of a workplace incident will be discussed.

C0-presenter Amanda Groller, MS.Ed., LPC will discuss the mental health impacts of a disaster.

589851 Advanced Modeling for Structured Packing Development - a: Hydraulics

Abstract:

Semi-empirical models such as the Rocha-Bravo-Fair, Delft, and Billet-Schultes models represent the current state-of-the-art in predicting the hydraulic and separation performance of structured packings. However, these models are derived from experimental data that cover a limited range of operating conditions and chemical systems, thus precluding them from being truly predictive. To develop a more robust and predictive hydraulic model for structured packings, computational fluid dynamic (CFD) modeling is utilized to provide greater insight into the multiphase flow physics present inside these packings. Specifically, this study focuses on the observations of transitional and laminar flow structures inside simplified Representative Elementary Units (REUs) of structured packing predicted via CFD modeling of industrially relevant chemical systems. In addition, the impact of these transitional and laminar flow observations on hydraulic model development is discussed. Finally, we present CFD results illustrating that de facto standard turbulence models used previously in the open literature to model flow physics inside structured packings may fail to capture transitional and laminar flow effects properly and thus potentially produce significant errors in predicted pressure drop values.

588824 Advanced Modeling for Structured Packing Development - B: Mass Transfer

Abstract:

Vapor-liquid contactors are ubiquitous in the chemical separations field and are commonly used for distillation, absorption, and stripping processes. Structured packings are the most advanced column internals due to their low pressure drops, good chemical separations, and high throughputs. However, the transport phenomena occurring inside these structured packings is not fully understood, limiting efforts to optimize the packings. Computational fluid dynamics (CFD) simulations are a promising method to reveal the physics occurring within the packings and support packing innovation. In particular, CFD can predict the microscale mass transfer occurring within the structured packings, allowing for prediction of the overall chemical separation. These simulations are often challenged by the presence of turbulence in the system, primarily within the gas phase. This presentation will discuss predictions of turbulent effects on mass transfer for single-phase pipe flow and mass transfer results for a wetted-wall column. The results from this study have direct application to structured packing CFD simulations.

Improved Feed Entry Improves Parting Box Hydraulics

By

Henry Z. Kister, Garry Jacobs, and Abraham A. Kister

Fluor, Aliso Viejo, CA, USA

Presented at the AIChE Distillation Symposium 2020
Virtual, August 17 – 21, 2020

Improved Feed Entry Improves Parting Box Hydraulics

By Henry Kister, Garry Jacobs, and Abraham A. Kister, Fluor, Aliso Viejo, CA

Achieving good liquid distribution is crucial for the performance of packed towers and multi-pass tray towers. With the focus on liquid distribution, one device that has received surprisingly little attention in the published literature is the parting box. While good-practice guidelines are available, the published literature is short of work on the hydraulics of parting boxes and tested guidelines for its design.

To gain insight into the parting box performance, we built a scaled-down model, geometrically similar to the parting box in a hydrocarbon tower, and tested it with water. This model was intended not only to provide insight into the tower issue, but also to provide generic insight into parting box hydraulics as tested in a realistic parting box design. The experiments conducted extended well beyond the specific design under investigation into generic parting box hydraulics.

Our earlier tests[1] showed the liquid in the test parting box was highly frothed up at both high and low liquid rates. The frothing was due to water jets from the feed pipe entraining air into the water upon impact. This generated excessive froth heights in the parting box, and premature overflow at high rates. The liquid split to the five clusters of holes at the floor of the parting box was good at the maximum rates but quickly deteriorated as the liquid rate was turned down.

Adding a Perforated Impingement Plate (PIP) with either 20% or 10% open area, mounted either 2" or 4" above the parting box floor, greatly improved the split of liquid to the clusters of holes at the floor of the parting box. With a PIP, the split was very good to excellent both at the maximum rates and at turndown down to about 50% of the maximum rates. Lower rates were not tested.

In the tests with PIP's, an air gap formed beneath the PIP, with tall froth above, producing premature overflows at high rates, and low and relatively less aerated froth beneath the PIP. The air gaps resulted from the water jets issuing from the feed pipe entraining air with them. This air descended through the PIP and accumulated beneath the PIP, being trapped by the prevailing water head on top of the PIP, and pressuring up the space between the parting box floor and the PIP.

As our previous tests modelled an existing parting box, the water to the parting box issued from a perforated sparger pipe at velocities of 6-7 ft/s at the maximum rates. While still within the good practice range[2] of less than 10 ft/s, these velocities exceeded the preferred range[3,4] of 4-5 ft/s.

Our current work addresses the question of whether bringing the hole velocities into the preferred range would mitigate the frothing. In the current series of tests we used two alternative pipe spargers with bigger holes, one at a time, which reduce the discharge velocities. This extends our hydraulic insights over a wider range of sparger hole velocities.

To avoid frothing, many parting box designs equip the pipe sparger perforations with dip tubes that bring the liquid below the liquid surface, often a short distance above the floor of the parting box, and thus avoid entrainment of vapor into the parting box liquid. This is a more expensive design, often considered ideal. Our current work examines the hydraulics of the parting box with dip tubes that bring the liquid into the box below the liquid level.

Test Rig and Procedure

The current tests were performed using the test rig reported in our previous work[1]. The only change was the feed piping. Our test rig is a scaled-down model, geometrically similar to the parting box in a hydrocarbon tower.

Figure 1 shows a top view of the test rig. The parting box was rectangular acrylic 21.5" long x 2.75" wide x 9.5" tall, and contained a perforated floor plate 20.5" x 1.75" cemented to the bottom flange. The floor plate had 75 holes at 3/16" diameter drilled into it using a press drill, in five clusters as shown in Figure 1, mimicking the hole geometry and arrangement in the floor of the hydrocarbon tower parting box. The total perforated area was 5.8% of the floor plate. The top of the rig was saddle-shaped to support the 2.07" ID transparent PVC perforated pipe sparger.

The peripheral ½" wide flanges surrounding the hole field of the floor plate were used for mounting two peripheral ½" wide hollow rectangular inserts, one 4" tall, the other 2" tall, one above the other. These inserts were used in some of the current tests. A flat momentum-breaking plate, referred to as a Perforated Impingement Plate (PIP), such as that used in many commercial parting boxes to arrest the impact of the liquid jets issuing from the sparger on the liquid surface in the parting box, was sandwiched between the inserts. Two PIP's were used in the tests, one at a time, each containing 3/16" perforations, one with 10% open area, the other with 20% open area. The inserts permitted each PIP to be mounted either 2" or 4" above the floor.

The sparger, shown in Figure 1 and used in our previous test series, had 12 downward-pointing 3/8" holes, six equally-spaced on each side of the tee. In the current series we tested two new feed pipes (one at a time), each with 12 downward-pointing holes in the same locations as in Figure 1, and with the bottom of the pipe 13" above the parting box floor. The first of the two had ½" holes, the second 5/8" holes. Round hose clamps were used to fix the pipe along the parting box length so the center lines of the tee lined up directly above the center lines of the floor plate. Two 5/32" vent holes, each at the opposite end of the tee, were drilled at the top of the pipe. Loose round hose clamps were slip-fitted around the pipe and moved over the vent holes to cover them after the air vented out at the beginning of each run.

Tap water to the rig was supplied by a 25 gpm submerged centrifugal pump from the bottom of a 96 gallon drum. A throttling valve in the pump discharge line to the sparger enabled manually adjusting the flow rate. For many runs, the valve was operated fully open and the pump delivered 25 gpm with a discharge pressure of 0.5 psi. For lower flow rates, the valve was set at a fixed throttle right at the start of the run. The water level on the drum was kept constant within plus or minus a foot by pumping water into it from two other drums, a 64 gallon drum and a 55 gallon drum, using two additional submerged pumps. In runs where more than 25 gpm were required, water to the rig was supplied using a 40 gpm centrifugal pump with valve throttling to the desired flow rate. The tests were conducted in the open air in dry weather with ambient temperatures ranging from about 65°F to 85°F.

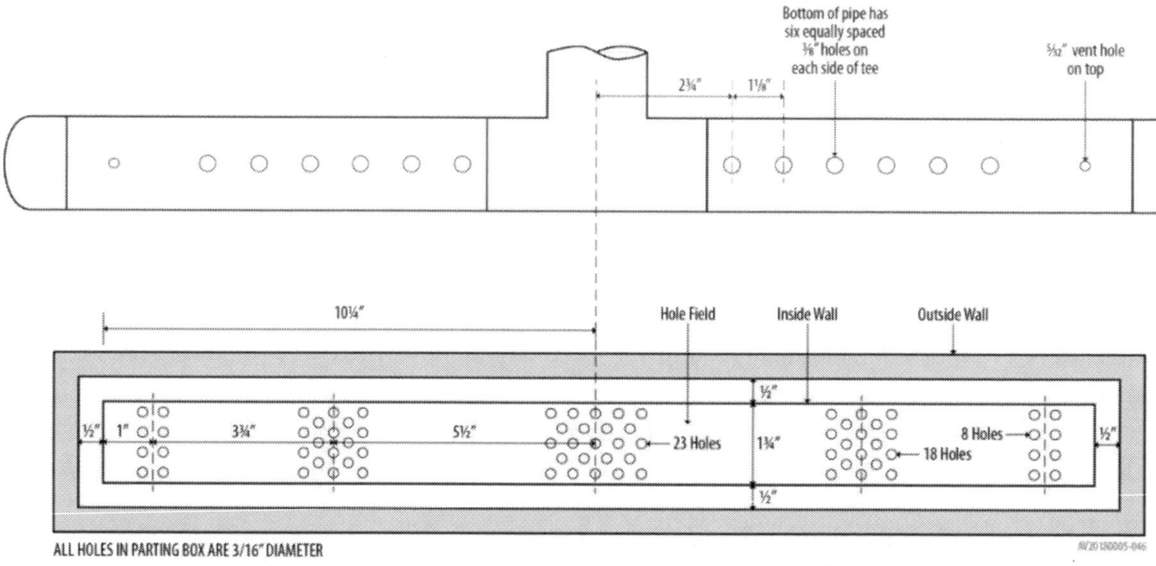

Figure 1. Test Rig Sketch: perforated pipe and parting box floor plans

Water leaving each cluster of holes in the floor plate descended into a separate chute, each diverting the collected water into a separate collector bin by gravity via a flexible pipe containing bellows. Each bin contained a ruler vertically fixed to the bin wall. At the end of each run a photograph was taken of the liquid level at the ruler in each bin. Prior to the tests, and again at the end of the tests, each bin was calibrated by stepwise introducing measured quantities of water and recording the water height in the bin. At the end of the tests the calibration was repeated, this time by stepwise removing the water from each bin by fixed quantities and again recording the heights in each bin after each step. The differences between the two calibrations were small in all the bins. Typically, in the larger bins collecting the liquid from the central three clusters, 1 inch of height corresponded to about 3 gallons, while in the smaller bins collecting liquid from the side clusters 1 inch of height coincided with about 2 gallons.

Each test was concluded after about 110-140 gallons were collected and the central bin approached its maximum desired height. Upon a signal, the pump was stopped and all the bellowed pipes were simultaneously turned upwards so the water remaining in the rig was trapped by the pipes and no longer entered the bins. After all the bin water height readings were completed, the water was pumped back to the feed drums for the next run. At the end of the tests, the water was pumped into garden irrigation hoses so it was all reused and none wasted.

Test Results: Different Hole Sizes, No Perforated Impingement Plate (PIP)

Table 1 presents key calculated hydraulic parameters for the feed spargers tests using different hole diameters. For the pipe with 3/8" perforations, at flow rates below 20 gpm, the pipe hole velocities were within the desired velocity guidelines. At 25-27 gpm, the velocities were above the desired values[1,3] but still within the recommended range[1,2]. The pipe hole velocities were well above the pipe velocity, per the recommended practice guidelines[4,5]. For the pipe with ½" perforations (green shading), the pipe hole velocities at all test rates were within the desired velocity guidelines[1,3]. The hole velocities were about 3 times higher than the pipe velocities, which is considered to provide good distribution of liquid out of perforated pipes[4,5] according to good practice guidelines in the literature. For the pipe with 5/8" perforations, the hole velocities were quite low, probably lower than usually seen in practice. Also, with hole velocities being only about twice the pipe velocities, there would be some tendency for the exiting liquid to form a maldistributed pattern as it exits the feed sparger pipes[4].

The calculated liquid head in the parting box, as determined by the orifice equation (with coefficients per Perry's Handbook[6]) was well below the overflow, as recommended. The only good practice guideline not adhered to was that the two middle pipe perforations of each lateral fell directly above the second and fourth floor hole clusters. This pattern resulted from mimicking the parting box in the actual tower under investigation.

Figure 2 shows photographs of the parting box at 3-4 different liquid rates with no PIP's. A separate set of photographs is presented for each pipe perforation diameter. Figure 2a shows that for the 3/8" holes, where the hole velocities were high relative to the pipe velocities, the liquid jets from the pipe holes were quite even and pointed straight down, with only a slight horizontal slant in some of the runs. This is to be expected when the hole velocity is well above the pipe velocity[4, 5]. The horizontal slant of the liquid jets issuing from the holes increased with the larger holes (Figure 2b, and more so, Figure 2c) where the hole velocities were lower relative to the pipe velocities, with the jets slanted towards the ends of the parting box.

Figure 2 shows intense frothing in the parting box at all the rates. The frothing occurred beneath the pipe holes, while liquid directly under the unperforated tee remained clear. Qualitatively, it can be argued that for a given water rate Figure 2 also shows a small reduction in frothing in the parting box when the hole velocities were largely reduced.

Figure 2 shows intense frothing in the parting box for all sparger pipe hole diameters at all the rates. The frothing occurred beneath the sparger pipe holes, while liquid directly under the unperforated tee remained clear. This liquid was very clear at all runs 20 gpm and less. At the higher rates (25-27 gpm) froth built up above the clear liquid in the 3/8" holes tests, but the froth buildup was much smaller in the ½" and 5/8" holes tests, as can be seen from comparing the top photographs in Figure 2a to those of Figs. 2b and 2c.

Figure 3 is a bar chart comparing the measured liquid split for these runs (various colors) with the expected split assuming every floor perforation delivered the same amount of liquid (blue bar). Figure 3 shows that for all pipe hole sizes, the liquid split from the parting box was good only for the 27 gpm run, where the froth height was high. The further the liquid rate was lowered, the more the liquid tended to issue via the middle cluster of holes, and the less through the off-center and/or side holes. One may argue that there was some improvement when the hole diameters were larger, but such an improvement appears small.

Table 1. Calculated Pipe and Hole Velocities

Liquid Rate, GPM	27.00	25.00	20.00	12.00	Recommended guidelines (see earlier)
Pipe velocity, ft/s	1.29	1.20	0.96	0.57	Hole velocity>>pipe velocity
Pipe rho vee squared, $lb/ft.s^2$	104	89	57	20	
Total number of pipe perforations	12	12	12	12	
Pipe perforation diameter, in	0.375	0.375	0.375	0.375	
Pipe perforation velocity, ft/s	6.54	6.05	4.84	2.90	<10 ft/s, desired <4-5 ft/s
Pipe perforation rho vee squared, $lb/ft.s^2$	2667	2286	1463	527	
Pipe perforation diameter, in	0.5	0.5	0.5	0.5	
Pipe perforation velocity, ft/s	3.68	3.40	2.72	1.63	<10 ft/s, desired <4-5 ft/s
Pipe perforation rho vee squared, $lb/ft.s^2$	844	723	463	166	
Pipe perforation diameter, in	0.625	0.625	0.625	0.625	
Pipe perforation velocity, ft/s	2.35	2.18	1.74	1.04	<10 ft/s, desired <4-5 ft/s
Pipe perforation rho vee squared, $lb/ft.s^2$	346	296	190	68	
Number of floor perforations	75	75	75	75	
Diameter of floor perforation, in	0.1875	0.1875	0.1875	0.1875	
Floor perforation velocity, ft/s	4.18	3.87	3.10	1.86	
Floor perforation rho vee squared, $lb/ft.s^2$	1092	936	599	216	
Calculated liquid head in parting box, inches	6.53	5.59	3.58	1.29	Avoid overflow (<9.5")

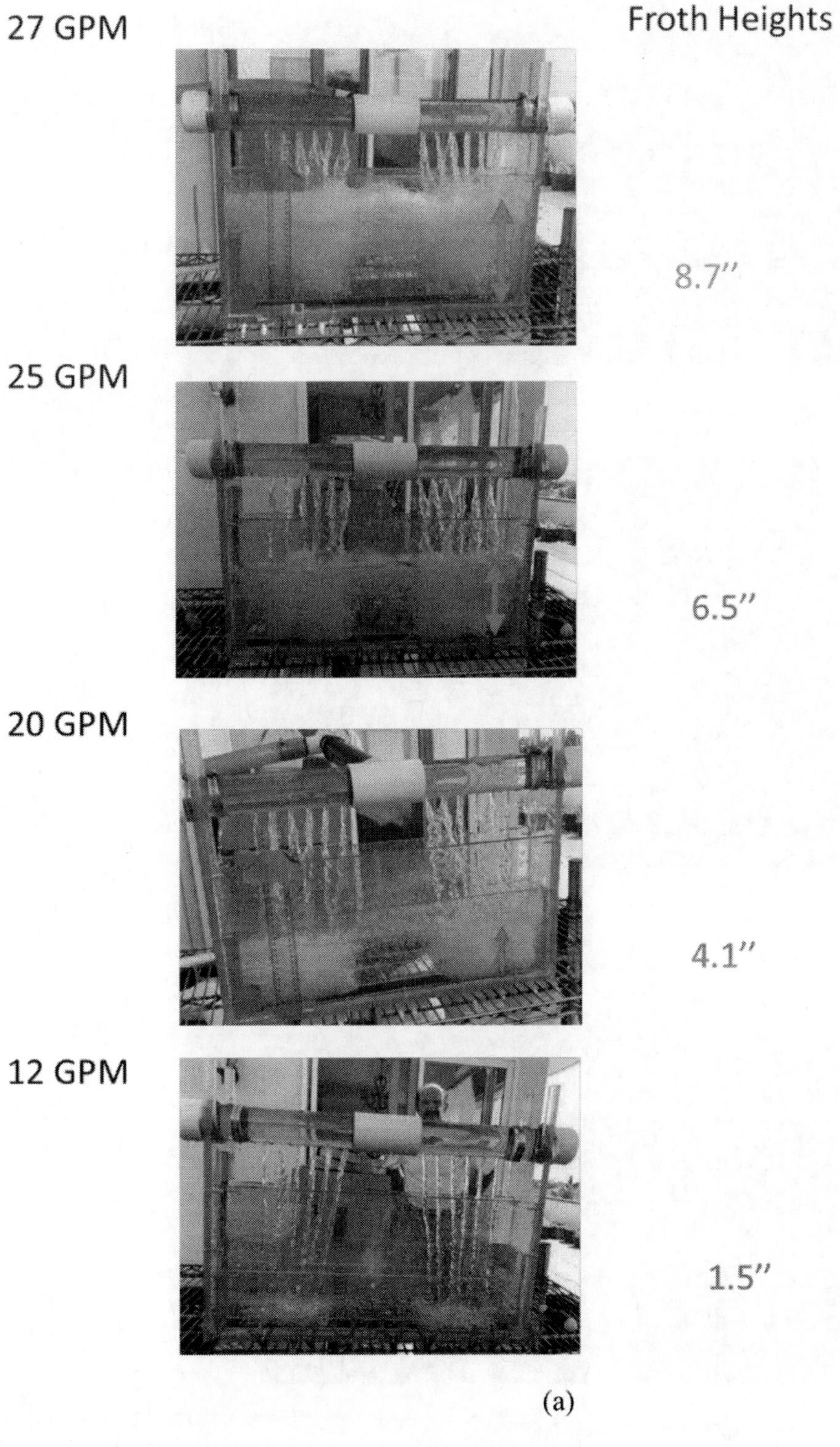

(a)

Figure 2. Tests with different size perforations in the feed pipes, no PIP.

(a) 3/8" holes; (b) ½" holes; (c) 5/8" holes

XA-Series ($^1/_2$ " Holes)

27 GPM

7.5"

18 GPM

3.0"

12 GPM

1.6"

(b)

YA-Series ($^5/_8$" Holes)

27 GPM

7.5"

18 GPM

3.0"

12 GPM

1.4"

(c)

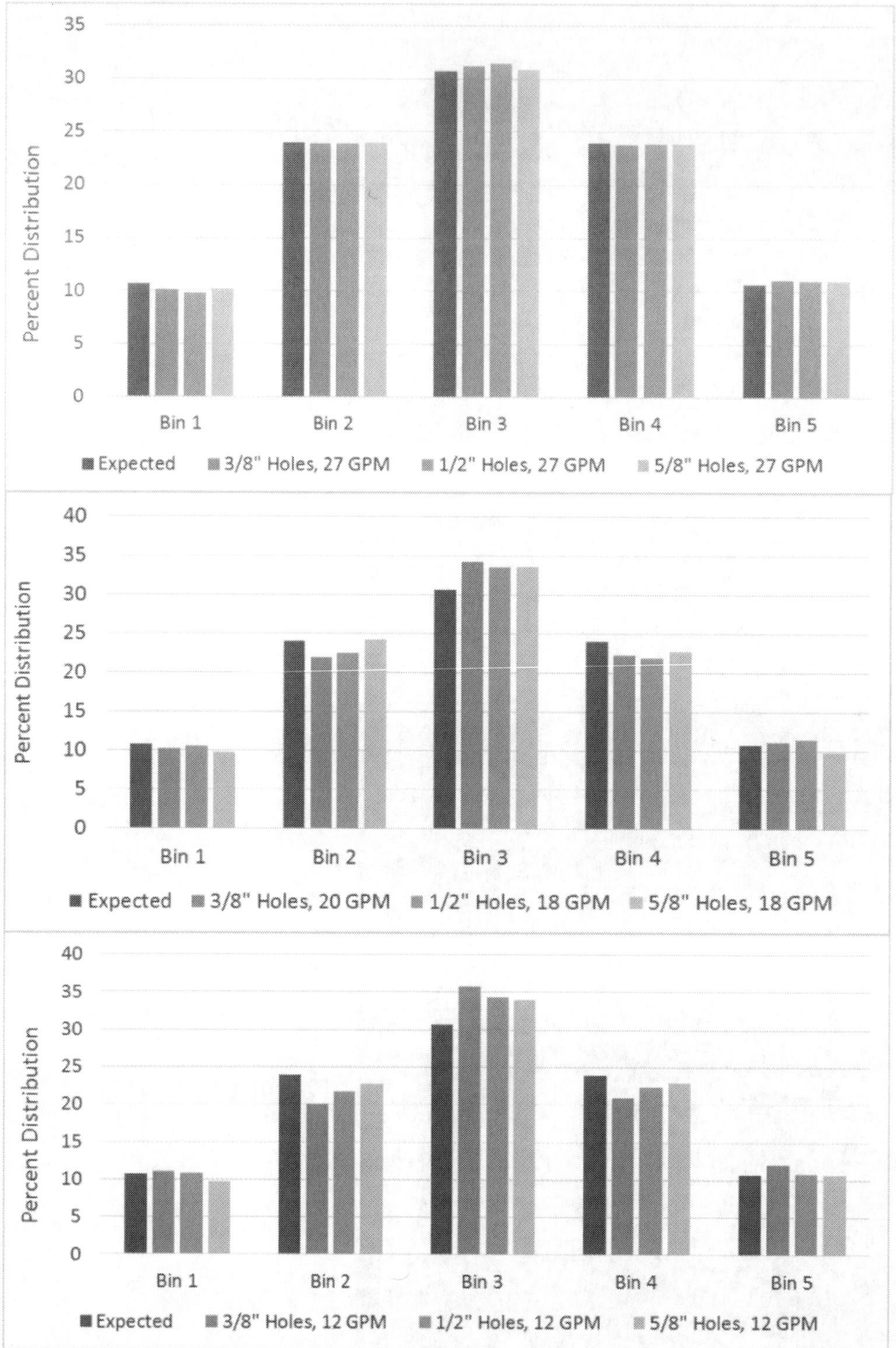

Figure 3. Liquid split among the bins compared to the desired split based of equal liquid flow rate through each cluster of perforations, no PIP tests.

Figure 4 compares the froth heights as determined visually from the photographs to the liquid heads calculated from the orifice equation (with coefficients per Perry's Handbook[6]). At liquid flow rates up to 20 gpm, the measured froth heights were slightly higher than the calculated liquid heads for all hole sizes. At 25-27 gpm, the measured froth heights for the 3/8 inch holes (hole velocities of 6-7 ft/s) rose to 20-30% above those calculated from the orifice equation. That rise was not observed for the larger holes. For the ½" and 5/8" holes, with hole velocities leaving the sparger of less than 4 ft/s, the froth heights remained very slightly above those calculated from the orifice equation.

At the lower rates, the liquid beneath the sparger tee was quite clear, whereas, below the liquid jets, the liquid was somewhat aerated. At the higher rates, with the high velocities out of the 3/8" holes, the aeration migrated into the liquid beneath the unperforated tee, but remained less frothy than on either side. Aeration reduces static head (i.e., less clear liquid), resulting in liquid maldistribution that favors the less aerated central region (Bin 3), relative to the peripheral regions. As the aeration increases with rate, so too does the froth height and the deviation from expected clear liquid head in Figure 4. At the lower rates, and with the low velocities issuing from the larger holes even at the higher rates (25-27 gpm), there was little froth migrating and piling up above the clear liquid (Figure 2), giving froth heights only slightly higher than the calculated liquid heads. The practical implication is that overflow is likely to occur earlier with high velocities out of the sparger holes.

Overall, the tests showed that reducing the sparger hole velocities slightly reduced the frothing in the parting box, which helped prevent overflow and gave froth heights close to the liquid heads calculated from the orifice equation. For all sparger hole sizes, the liquid split from the parting box was good only for the 27 gpm run, where the froth heights were high. The farther the liquid rate was lowered, the more the liquid tended to issue via the middle cluster of holes, and the less through the off-center and/or side holes. One may argue that there was some improvement when the hole diameters were larger, but such an improvement was small.

Test Results: Different Hole Sizes, with PIP's

We performed a limited number of tests with larger diameter holes with the 20% PIP's, all at the maximum liquid rate of 27 gpm, at elevations 4" and 2" above the floor. It is important to keep in mind that in comparison, the perforated area of the parting box floor plate is only 5.8% of the floor area. Figure 5 and 6 show photographs from the tests. Figure 5 shows that with the larger hole diameters (lower hole velocities) the froth heights were somewhat lower. This is less clearly seen also in Figure 6. The explanation for the lower froth heights at the lower pipe hole velocities is similar to that given for the no-PIP tests above. The migration of aeration into the region under the tee with the high-velocity jets issuing from the 3/8" holes is larger than with the lower velocity jets issuing from the ½" and 5/8" holes, as can be seen in Figures 5 and 6. With the 3/8" holes, this middle region was filled with froth which raised the froth height in the parting box.

As in previous tests, the split to the bins was again excellent with the PIP's at 27 gpm. No turndown tests were conducted with the PIP's and larger holes.

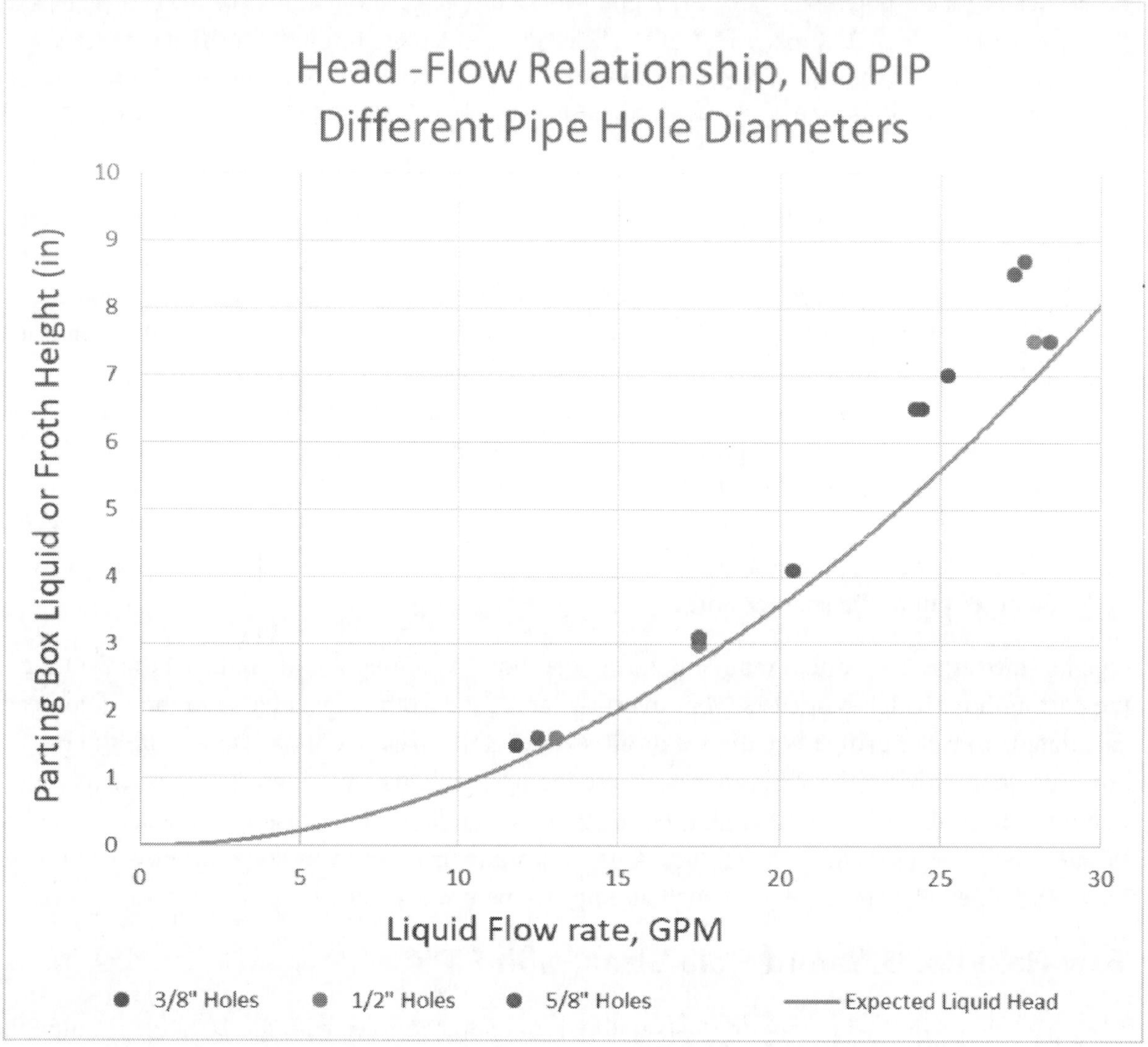

Figure 4. Head- flow relationship, Different Pipe Hole Diameters, No PIP tests.

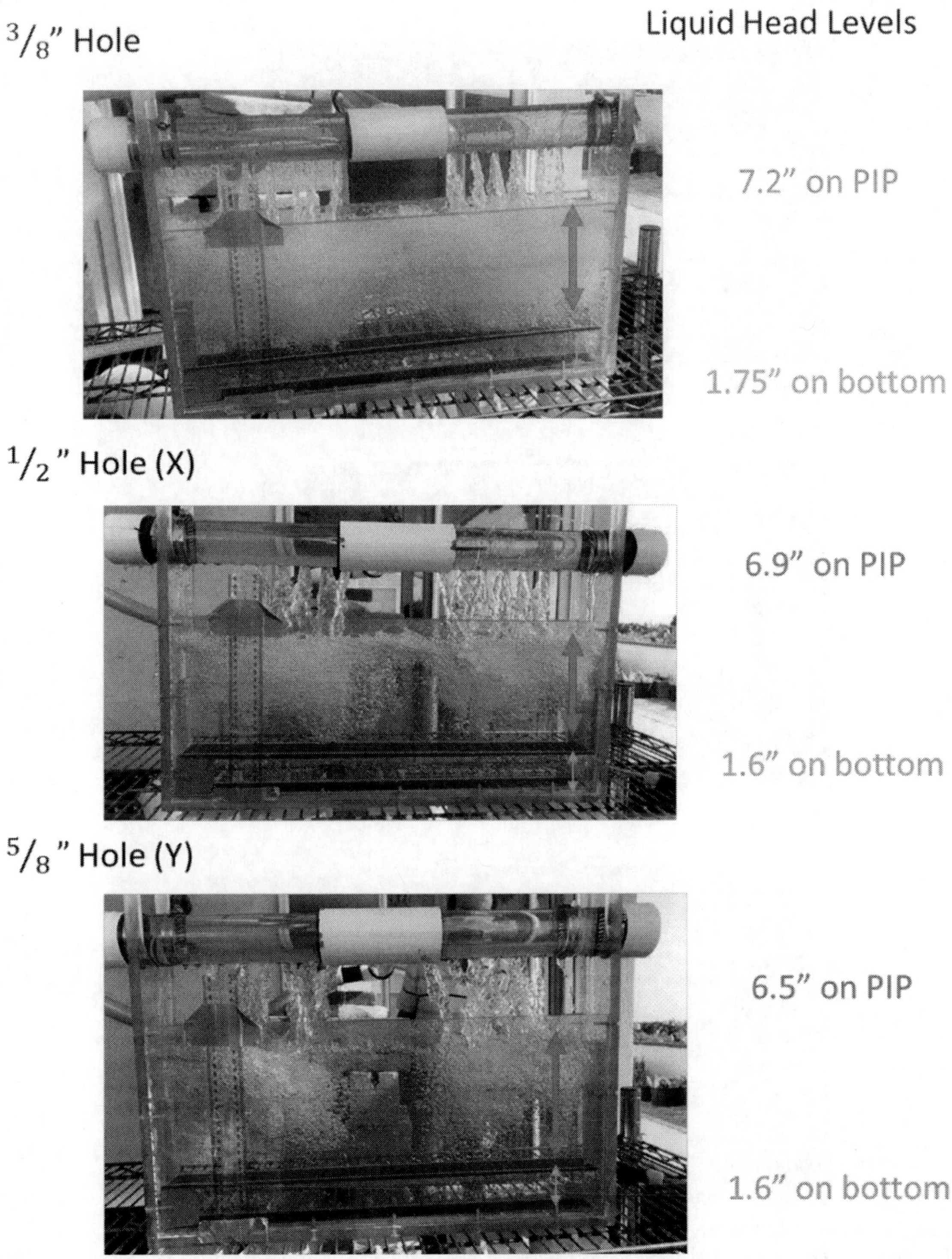

Figure 5. Tests at 27 gpm with 20% open area PIP, 2" above the parting box floor, with different feed sparger pipe hole diameters.

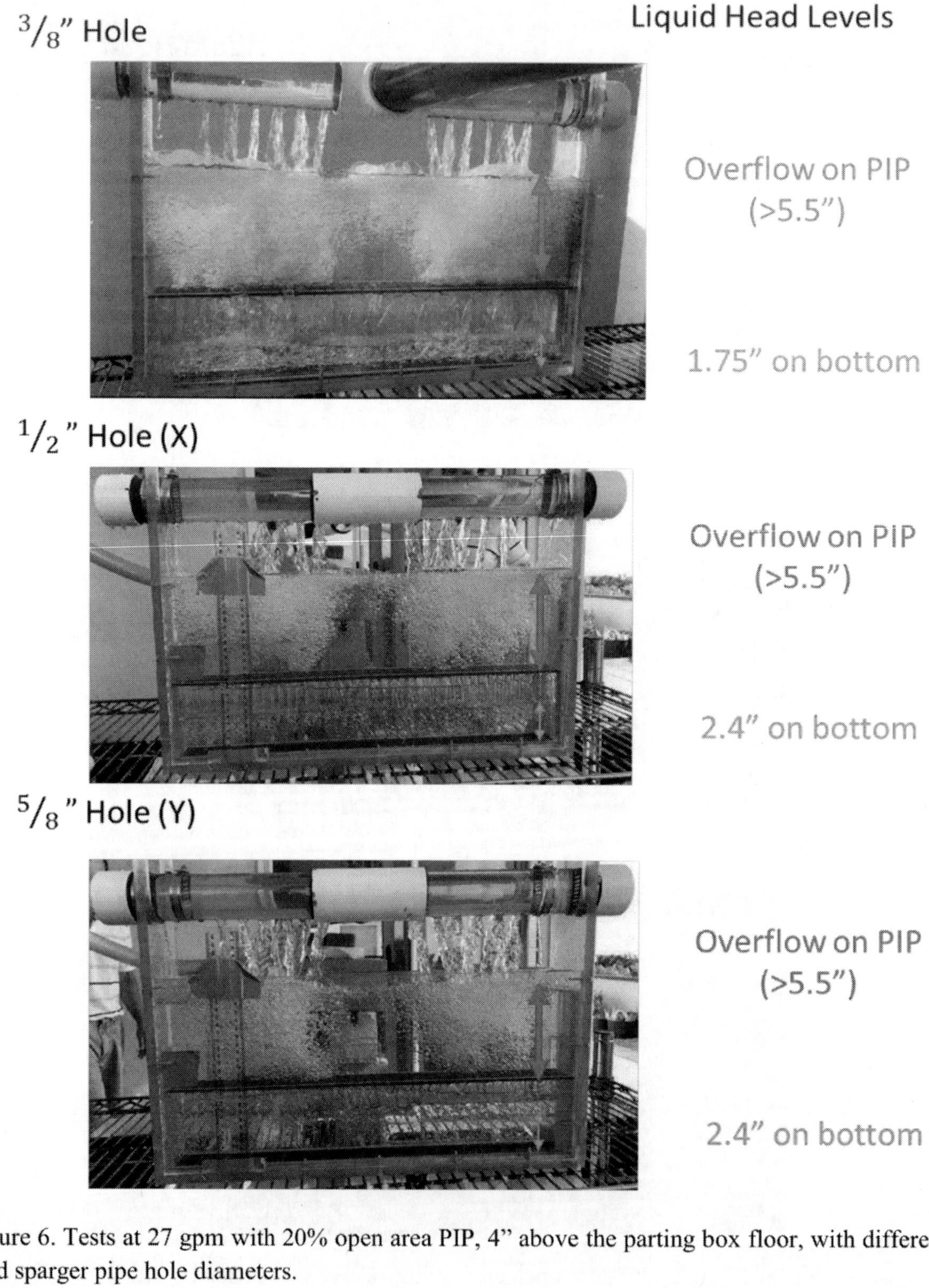

Figure 6. Tests at 27 gpm with 20% open area PIP, 4" above the parting box floor, with different feed sparger pipe hole diameters.

Tests Procedure with Dip tubes

To avoid frothing via vapor entrainment, many parting box designs equip the perforations with dip tubes (or guide tubes) that bring the liquid below the liquid surface, often terminating a short distance above the floor of the parting box. This is a more expensive design, often considered ideal. Our next series of tests examined the hydraulics of this practice in our model parting box.

Our tests were performed with four 0.622" ID tubes, inserted into the sparger feed pipe without restriction (i.e., full ID). Each dip tube discharged the liquid midway between every two adjacent clusters of holes. The diameter of 0.622" was selected to make the tubes roughly equivalent in cross section areas to the hole areas in the 3/8" holes tests, which were our base case. In the initial tests the tubes terminated about 1.75 inches above the parting box floor. In a later series of tests the tubes were shortened so that they terminated 5.5" above the parting box floor. This is to address another practice of using short directional guide tubes to prevent the horizontal projection of the liquid issuing out of pipe holes like in Figures 2b and 2c. The short tube practice also permits incorporating PIP's.

A final set of tests was performed with each dip tube terminating in a short tee that directed the liquid towards the parting box walls instead of the parting box floor. The tests with the tees at the ends of the tubes were repeated with the shortened tubes.

The photographs below show the tube arrangement. The tube velocities are compared to the velocities in the perforated pipe tests with 3/8" inch holes in Table 2. The tube velocities were 8% lower than the hole velocities with the 3/8" holes.

Table 2. Calculated Pipe, Hole, and Dip Tube Velocities

Liquid Rate, GPM	27.00	25.00	20.00	12.00	Recommended guidelines (see earlier)
Pipe velocity, ft/s	1.29	1.20	0.96	0.57	Hole velocity>>pipe velocity
Pipe rho vee squared, $lb/ft.s^2$	104	89	57	20	
Total number of pipe perforations	12	12	12	12	
Pipe perforation diameter, in	0.375	0.375	0.375	0.375	
Pipe perforation velocity, ft/s	6.54	6.05	4.84	2.90	<10 ft/s, desired <4-5 ft/s
Pipe perforation rho vee squared, $lb/ft.s^2$	2667	2286	1463	527	
Total number of dip tubes	4	4	4	4	
Dip tube diameter, in	0.622	0.622	0.622	0.622	
Dip tube velocity, ft/s	7.13	6.60	5.28	3.17	<10 ft/s, desired <4-5 ft/s
Dip tube rho vee squared, $lb/ft.s^2$	3171	2719	1740	626	

Test Results with Dip Tubes Terminating 1.75" above Floor

Figure 7 shows photographs of the tests with the dip tubes terminating 1.75" above the parting box floor. There was very little frothing in all except for the 14 gpm run when the parting box liquid level fell to near the bottom of the dip tube. At the 14 gpm test, the sparger feed pipe was not liquid-full, and air from the sparger feed pipe could be seen sucked into the dip tubes. We had the vent holes closed at this run but not tightly, so it appears that the descending water column in the tubes sucked some air into the feed pipe. At higher flow rates the feed pipe was liquid-full. This partly liquid-full sparger scenario with tower vapor sucked through vent holes is likely to also exist in commercial spargers with dip tubes.

Figure 8 shows a good, but not excellent, split of the liquid to the bins. In all runs, more liquid issued from the middle cluster and less through the sides and/or off-center clusters. There was no clear trend with the liquid rates.

Figure 9 compares the liquid heights as determined visually from the photographs to those calculated from the orifice equation (with coefficients per Perry's Handbook[6]). At liquid flow rates 20 gpm and higher, the measured liquid heights were about 30% higher than the calculated liquid heads predicted by the orifice equation. That rise was not observed for the 14 gpm run. For the liquid rates exceeding 14 gpm, frothing at the parting box was non-existent, so aeration could not explain the difference between the measurement and calculation. The most likely explanation is that the horizontal momentum of liquid issuing near the parting box floor made it more difficult for liquid to descend through the parting box holes, thereby incurring the additional liquid head. This led to a premature overflow.

Based on this explanation, and in an attempt to reduce the liquid head, we added tees to the bottom of each dip pipe, directing the liquid towards the parting box walls instead of to the floor. The ID of the tee was 0.840", which is the OD of ½" pipe. Figure 10 shows photographs from tests with the tees added. Compared to Figure 7, the liquid heads were lower, with Figure 9 showing they were well-predicted by the orifice equation.

Figure 11 shows an excellent split of the liquid to the bins. Figure 12 shows that this split is superior to that obtained without the tees.

Liquid Levels

27 GPM 8.5”

25 GPM 7.9”

20 GPM 4.8”

14 GPM 2.0”

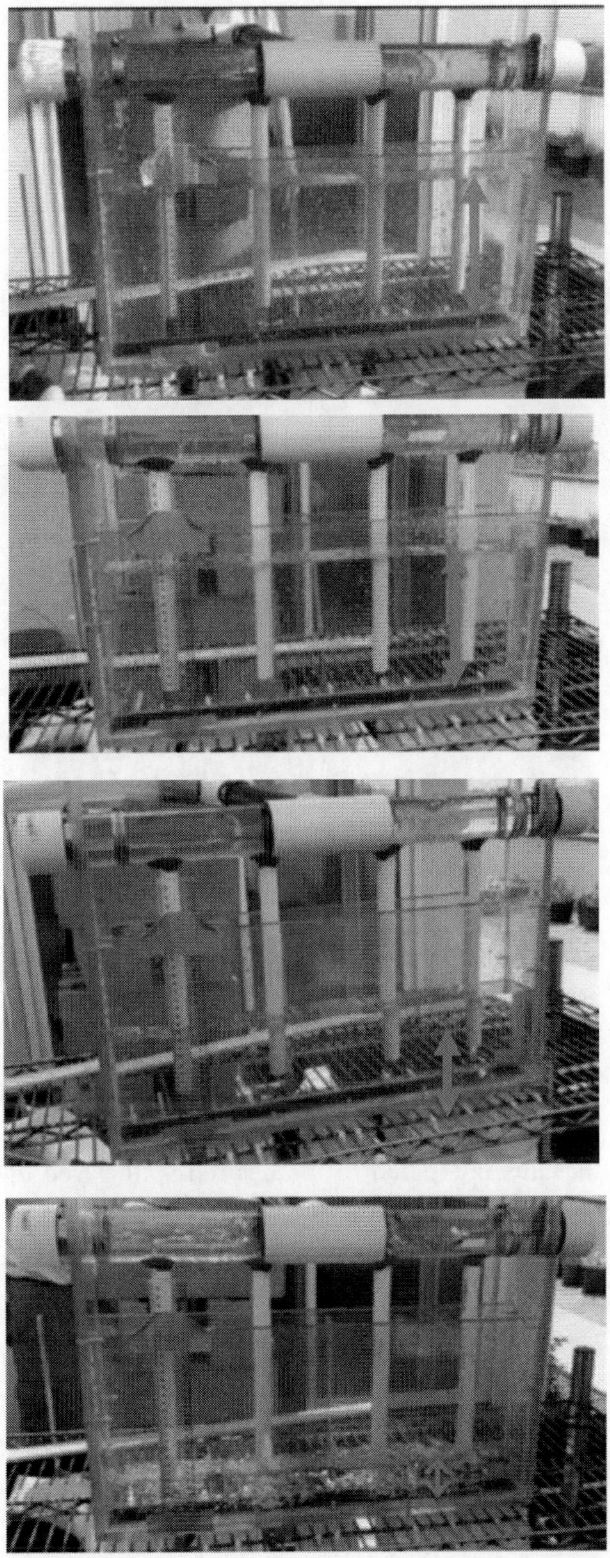

Figure 7. Tests with dip tubes terminating 1.75” above the parting box floor, no tees

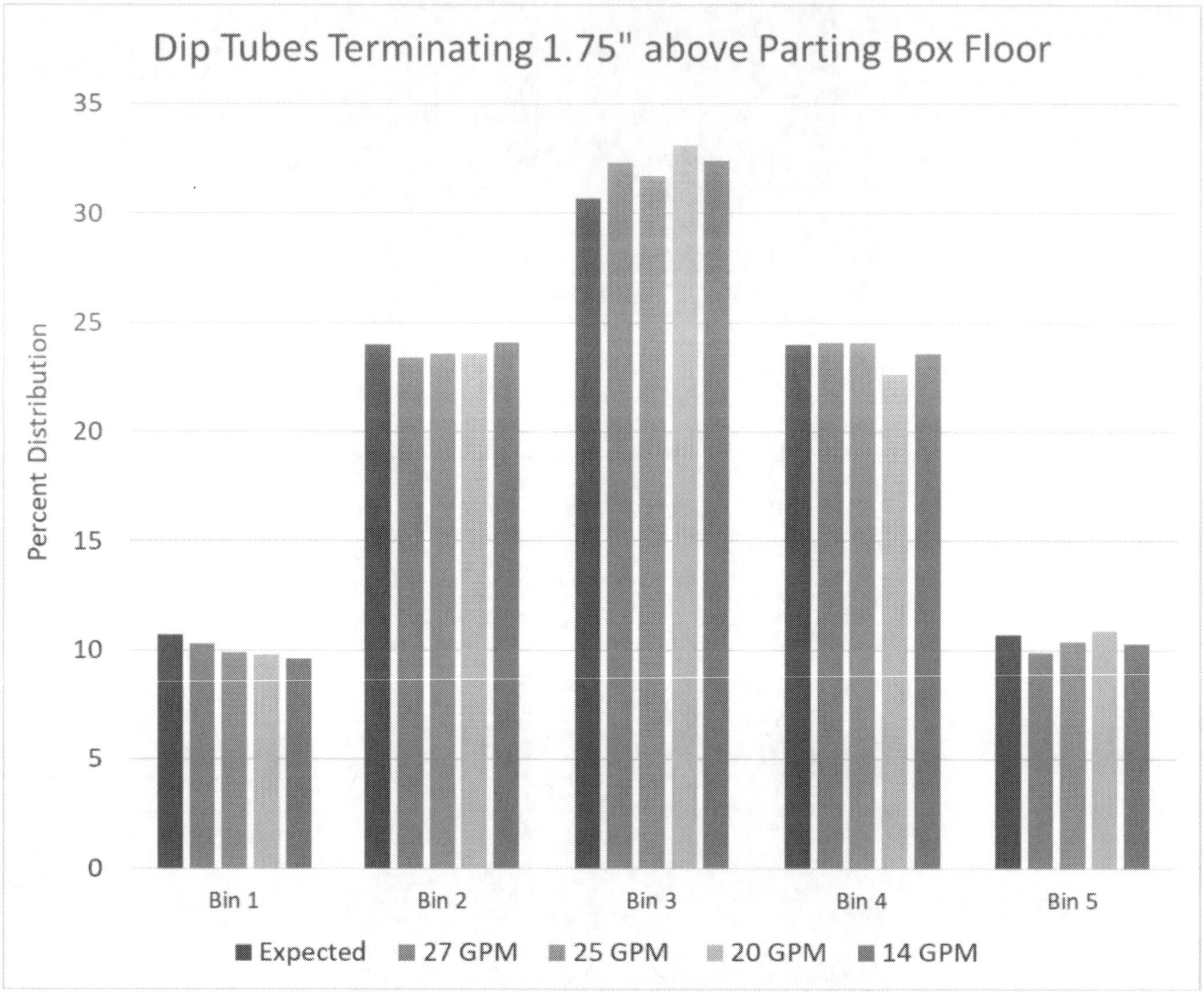

Figure 8. Liquid split among the bins compared to the desired split based of equal liquid flow rate through each perforation. Dip tubes terminating 1.75" above the parting box floor, no tees

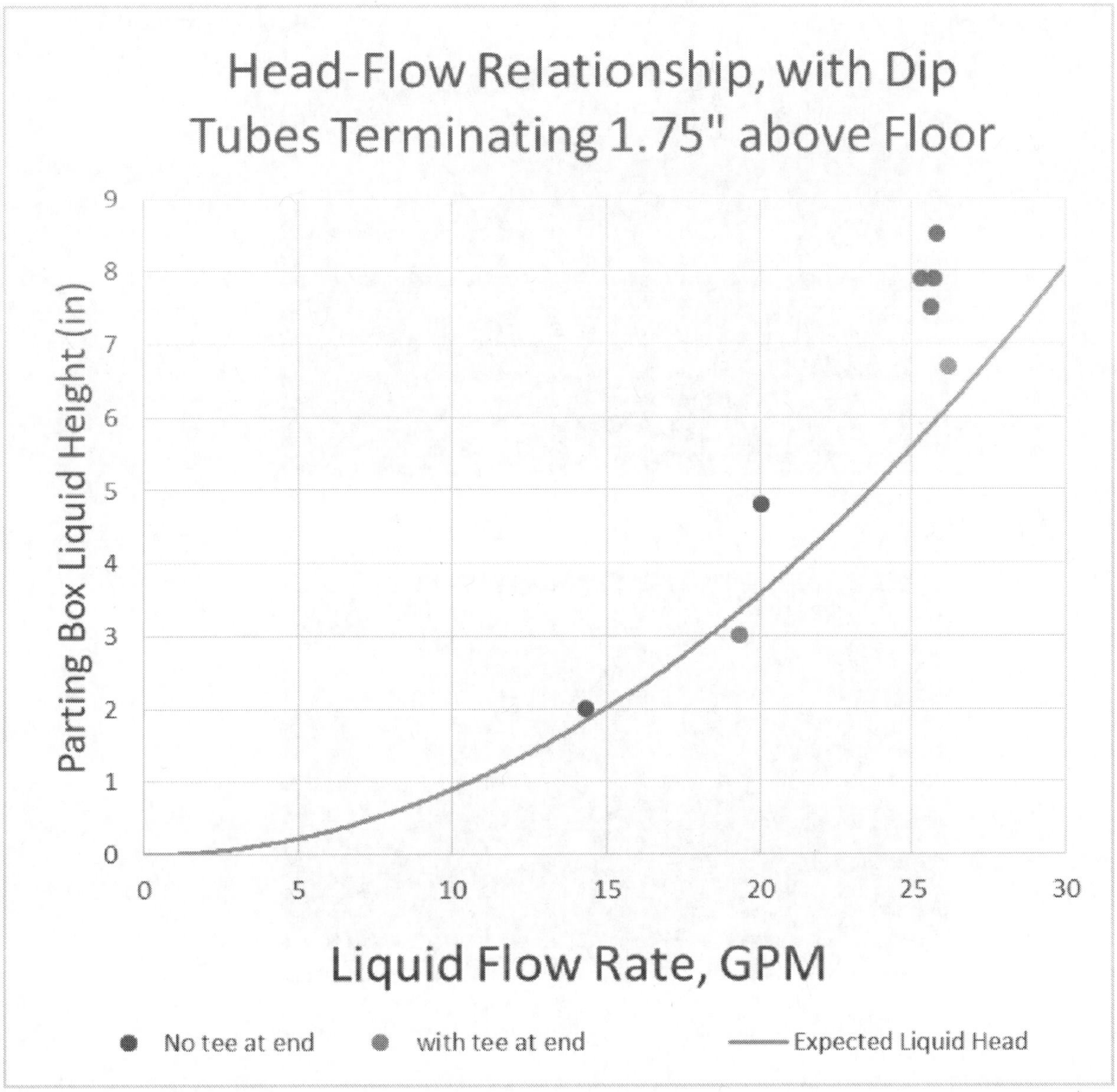

Figure 9. Head- flow relationship, with dip tubes 1.75" above the parting box floor, with or without tee at ends.

27 GPM

Liquid Levels

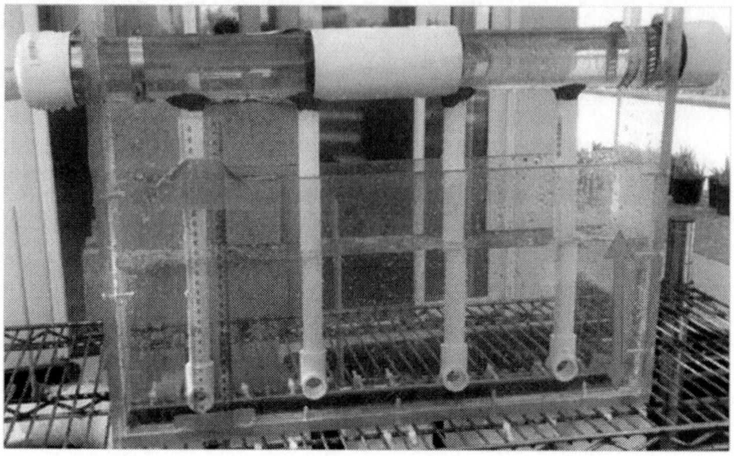

6.7”

20 GPM

3”

Figure 10. Tests with dip tubes terminating 1.75” above the parting box floor, with tees fitted at their bottom to direct the liquid towards the parting box walls

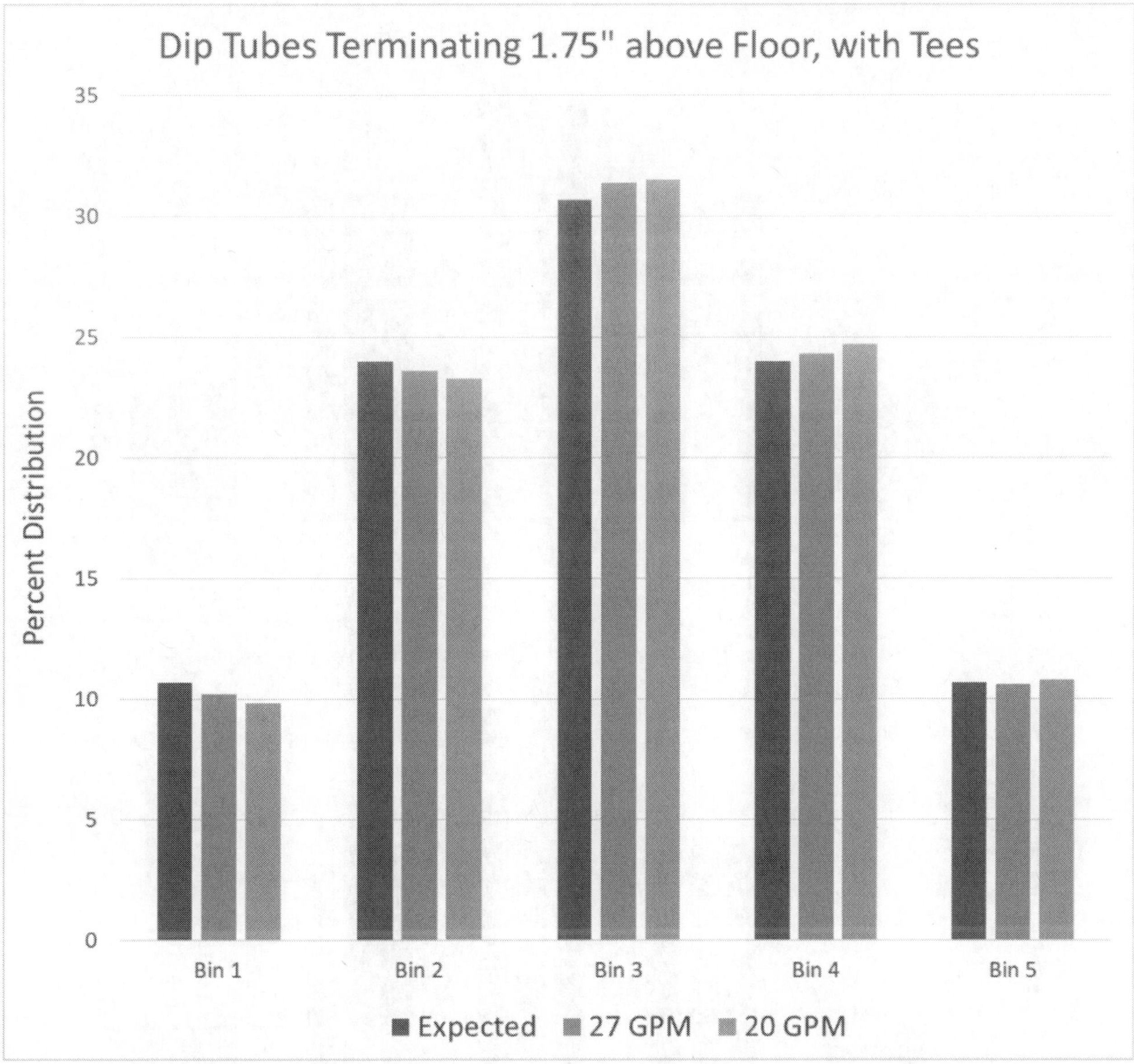

Figure 11. Liquid split among the bins compared to the desired split based of equal liquid flow rate through each perforation. Dip tubes terminating 1.75" above the parting box floor, fitted with a tee directed towards the parting box wall

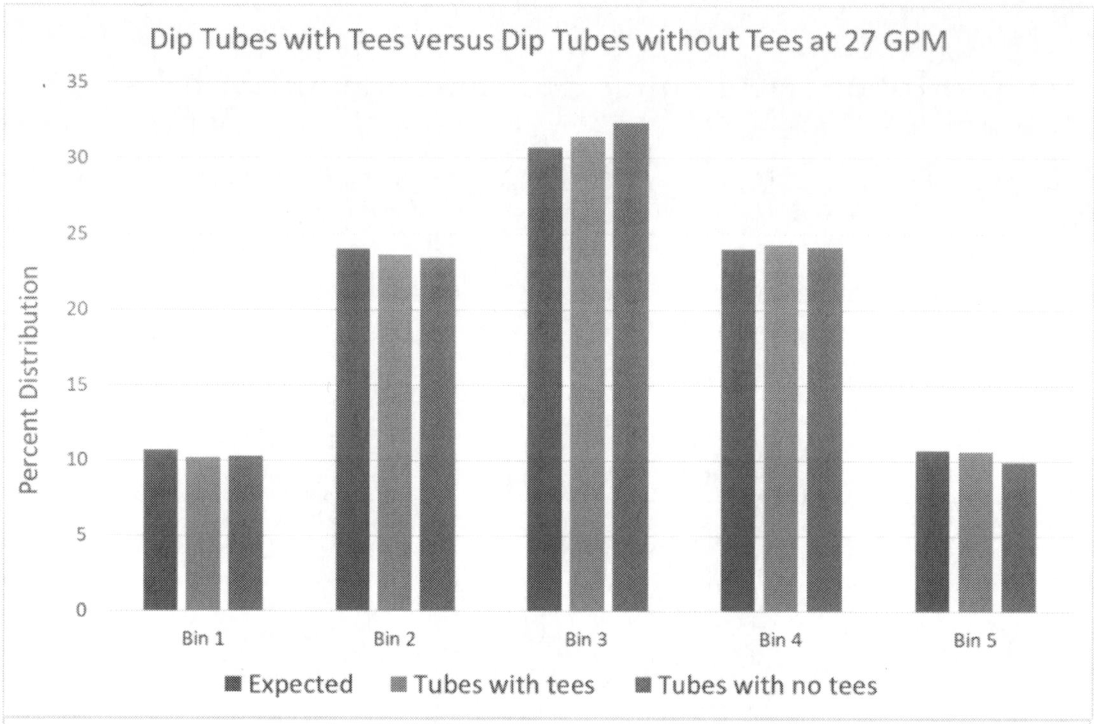

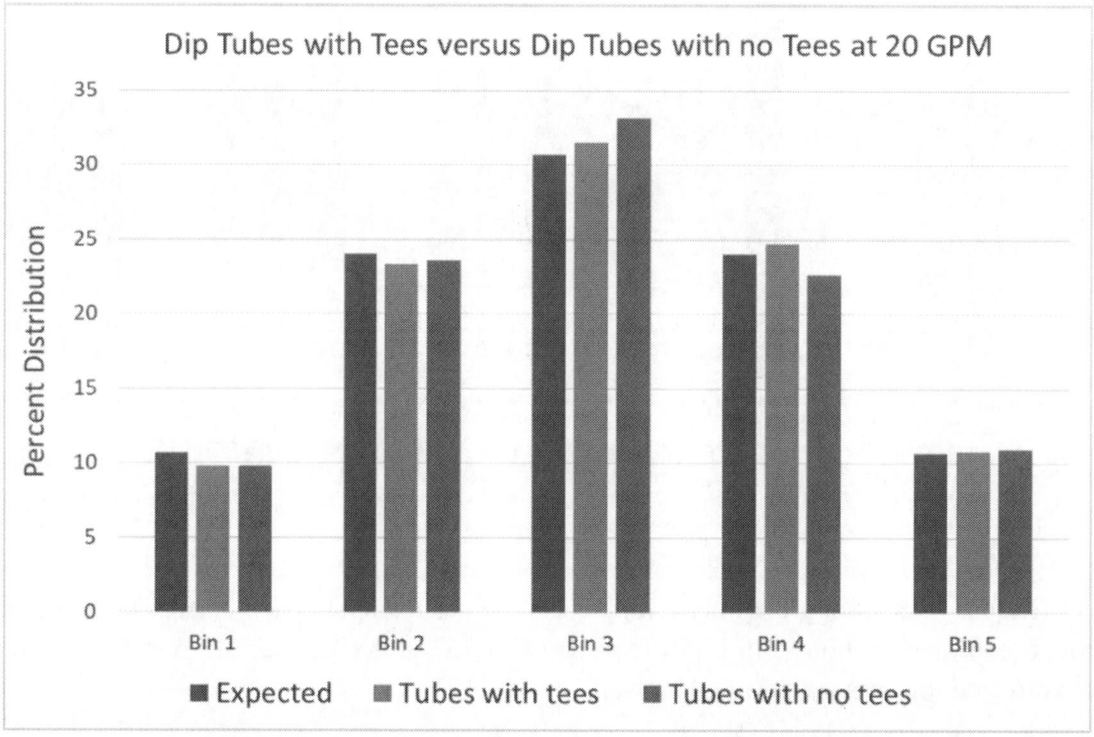

Figure 12. Liquid split among the bins compared to the desired split based of equal liquid flow rate through each perforation. Dip tubes terminating 1.75" above the parting box floor, compared to same dip tubes fitted with a tee directed towards the parting box wall

Test Results with Dip Tubes Terminating 5.5" above the Floor

Figure 13 shows photographs of the tests with the dip tubes terminating 5.5" above the parting box floor, part a with the tubes discharging liquid downwards, and part b with the tubes fitted with tees at their ends to direct the liquid towards the parting box walls. There was no frothing in the 27 gpm runs when the tubes were submerged in the parting box liquid, and some in the 12 gpm runs when the parting box liquid level was below the bottom of the tubes. Similar to the tests with the longer tubes, at the 12 gpm test, the feed sparger pipe was not liquid-full, and air from the feed sparger pipe could be seen sucked into the dip tubes.

Figure 14 shows an excellent split of the liquid to the bins with the tees fitted, and a good but inferior split without the tees.

Figure 15 compares the liquid heights as determined visually from the photographs to the liquid heads calculated from the orifice equation (with coefficients per Perry's Handbook[6]). Without the tees, the measured liquid heights were again about 30% higher than the liquid heads calculated from the orifice equation. Again, the most likely explanation is that the horizontal momentum of liquid jets hitting the parting box floor made it more difficult for the liquid to descend through the parting box holes, thereby incurring the additional liquid head. With the tees, the measured liquid heads were well-predicted by the orifice equation.

One more test was conducted with the dip tubes terminating 5.5 inches above the parting box floor. This was a repeat of the first test (no tee) but this time with the 10% PIP inserted 4 inches above the parting box floor, so the bottom of the dip tubes were approximately 1.5" above the PIP. Photographs are shown in Figure 16. At 25 gpm, the liquid height in the parting box was about 7.7" which submerged the dip tubes by more than 2 inches. There was no aeration in the parting box and no air gap under the PIP. As previously discussed, the aeration and air gap were caused by the impact of the liquid jets on the liquid surface, and since the tubes were submerged, there was no aeration and no air gap in this run. The frothing and air gap returned in the 12 gpm test, where the level of liquid in the parting box fell below the bottom of the tubes. Here there was impact, frothing, and air gap.

As with our previous tests with the PIP's[1], when there was frothing (the 12 gpm test) the froth height was considerably higher than predicted by the orifice equation, as shown in Figure 17. Two solid curves are shown in Figure 17. The lower (orange) curve is the same as in Figures 15, based on the floor orifices alone. The upper (purple) curve also includes the liquid head calculated across the PIP. Since the PIP and the floor open areas are 10% and 5.84%, respectively, of the parting box cross section area, the additional liquid head incurred by the PIP is significant.

Figure 17 shows that with the frothing eliminated (25 gpm), and no air gap, the measured liquid height with the PIP was the same as that calculated from the orifice equation (taking the PIP into account). Finally, Figure 18 shows that with the shortened tubes and the 10% open PIP the liquid split to the clusters was excellent both at the 25 gpm and the 12 gpm liquid rates.

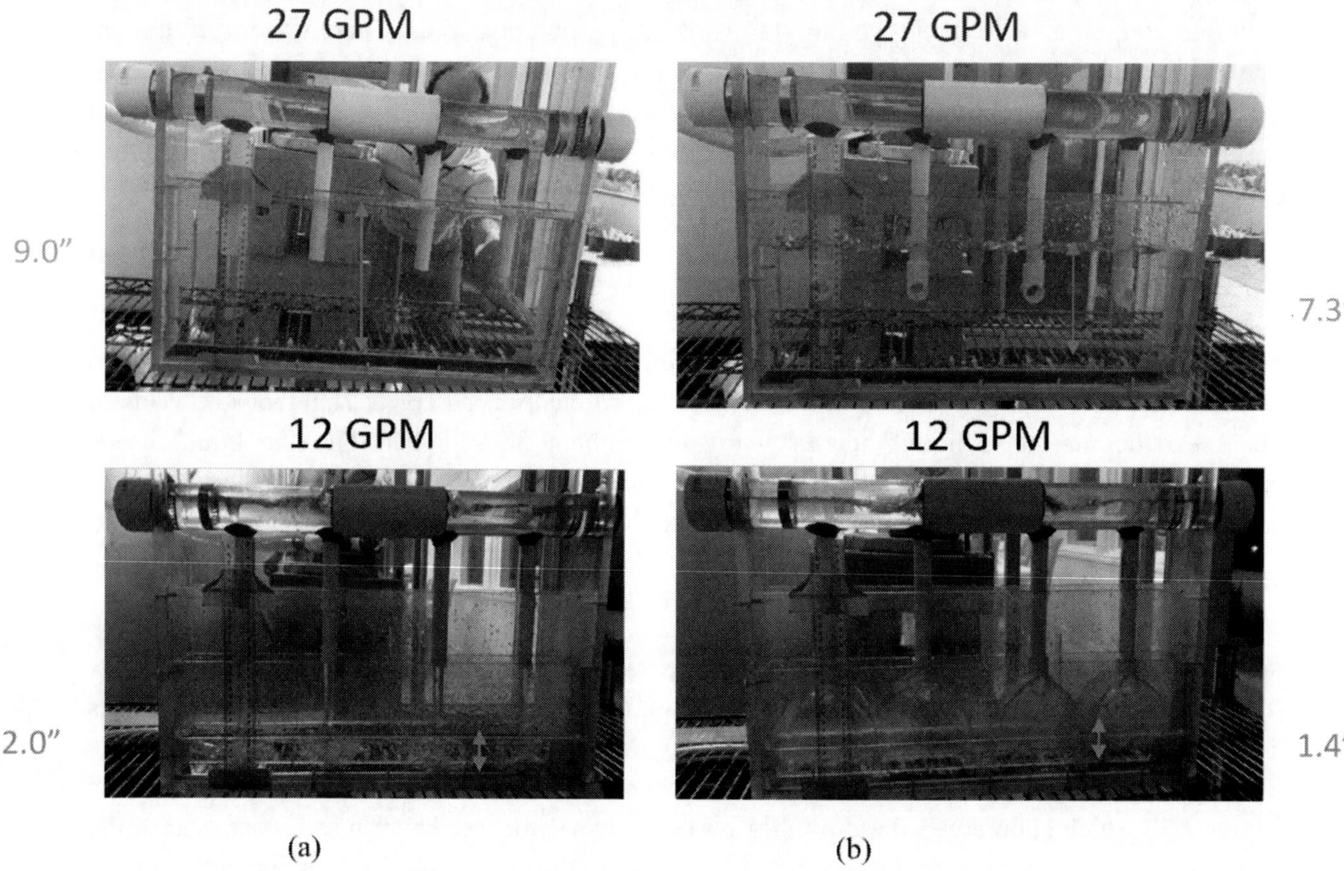

Figure 13. Tests with dip tubes terminating 5.5" above the parting box floor, (a) with liquid issuing downwards, (b) with tees fitted at their bottom to direct the liquid towards the parting box walls

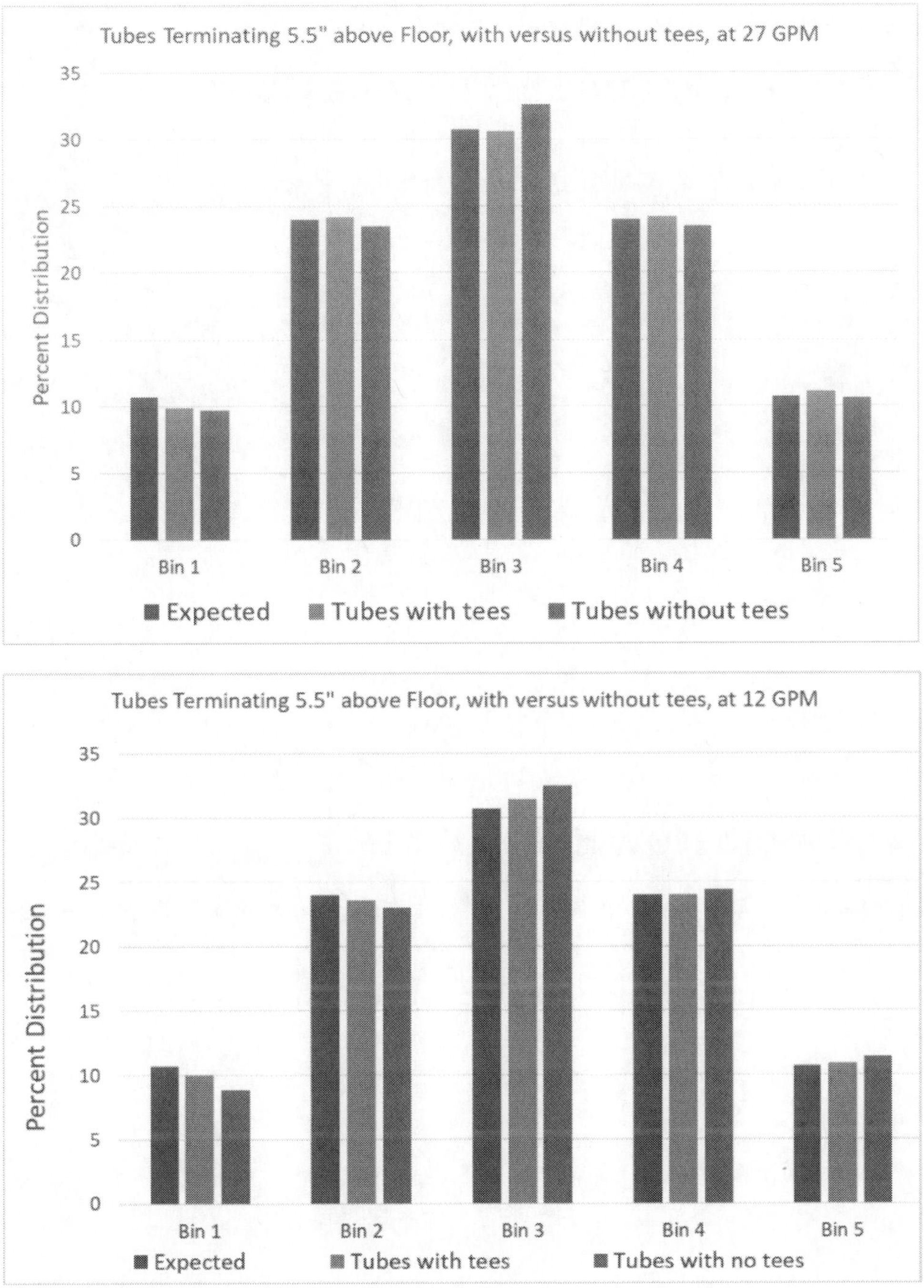

Figure 14. Liquid split among the bins compared to the desired split based of equal liquid flow rate through each perforation. Dip tubes terminating 5.5" above the parting box floor, compared to same dip tubes fitted with a tee directed towards the parting box wall

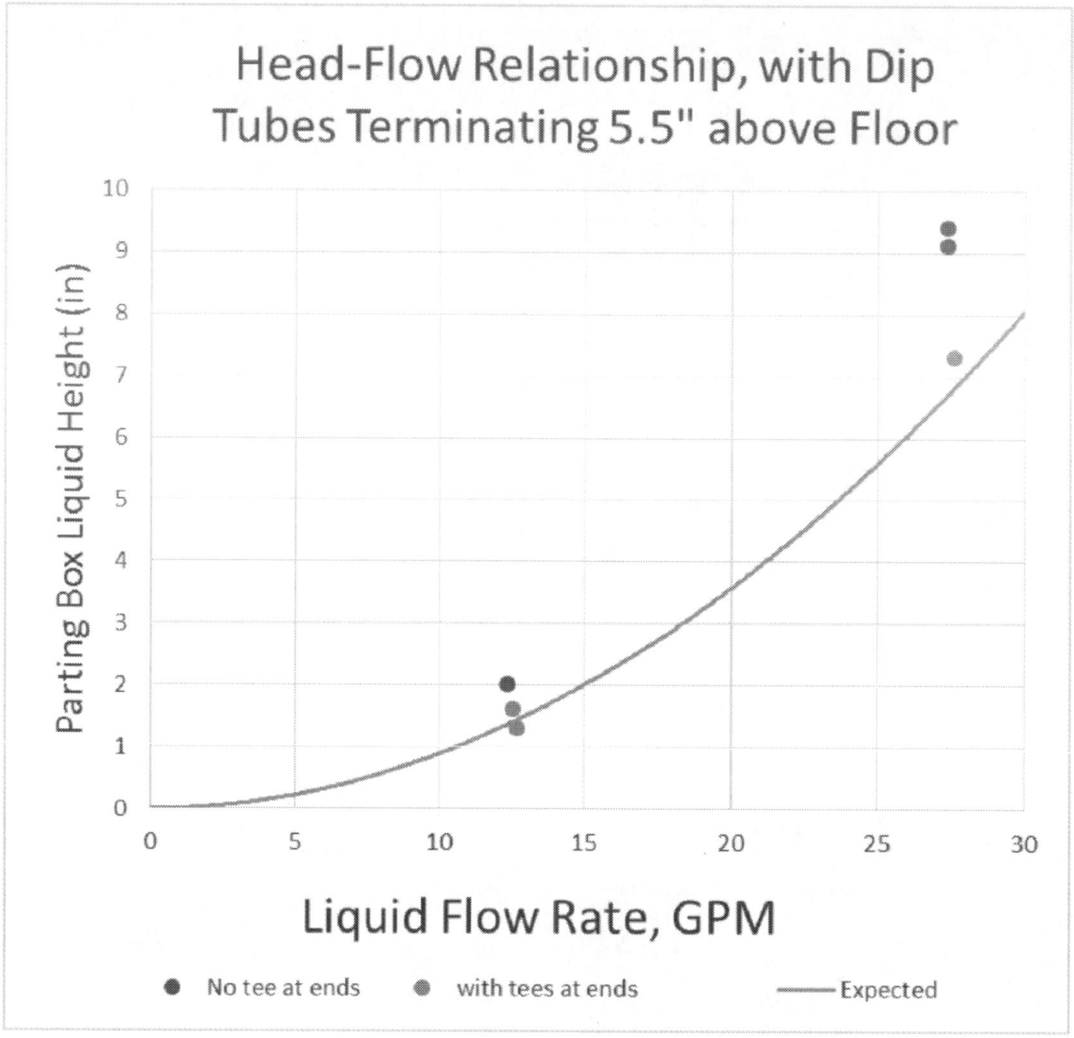

Figure 15. Head- flow relationship, with dip tubes 5.5" above the parting box floor, with or without tee at ends.

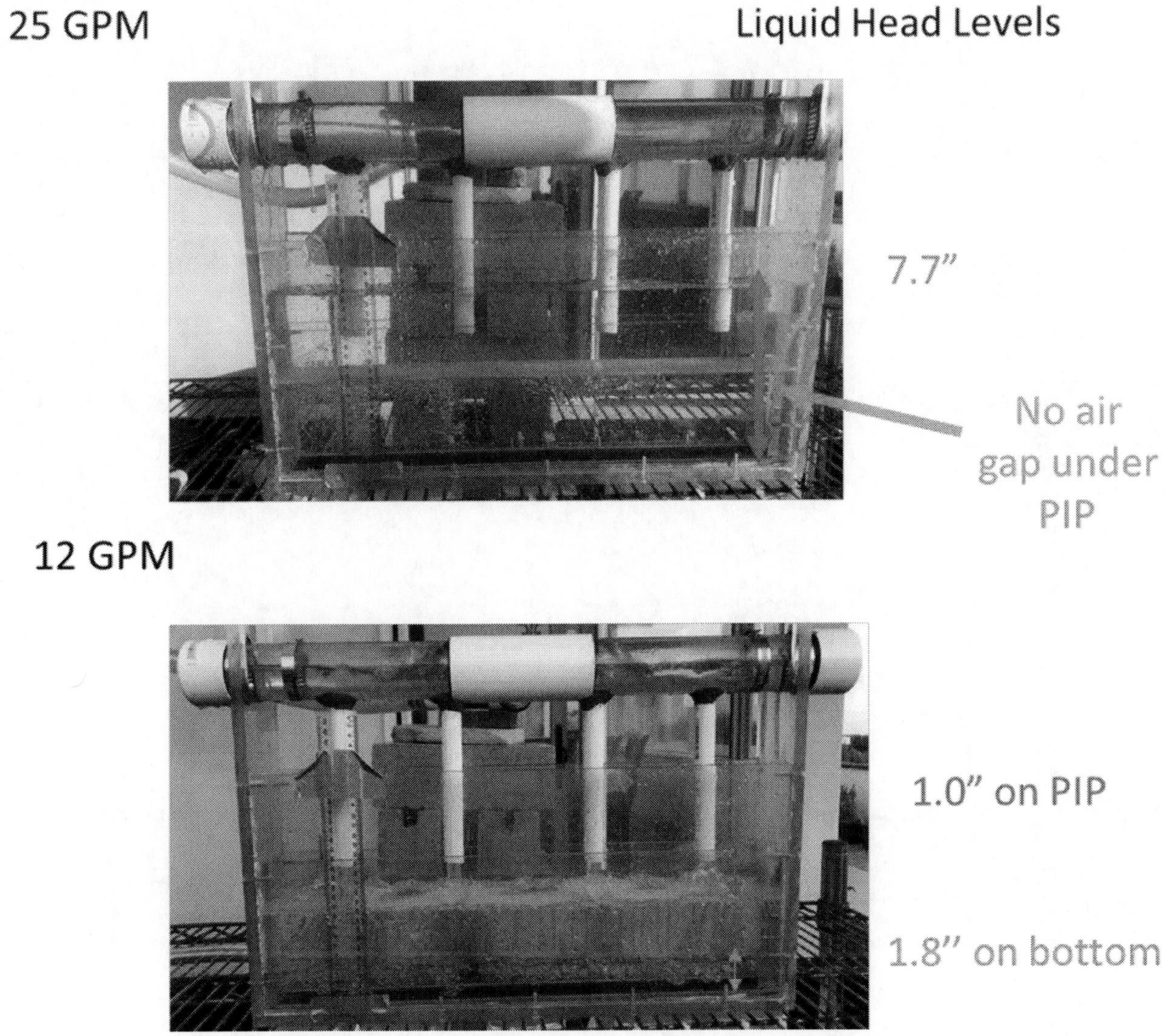

Figure 16. Tests with dip tubes terminating 5.5" above the parting box floor, with liquid issuing downwards (no tee), with the 10% PIP mounted 4" above the parting box floor.

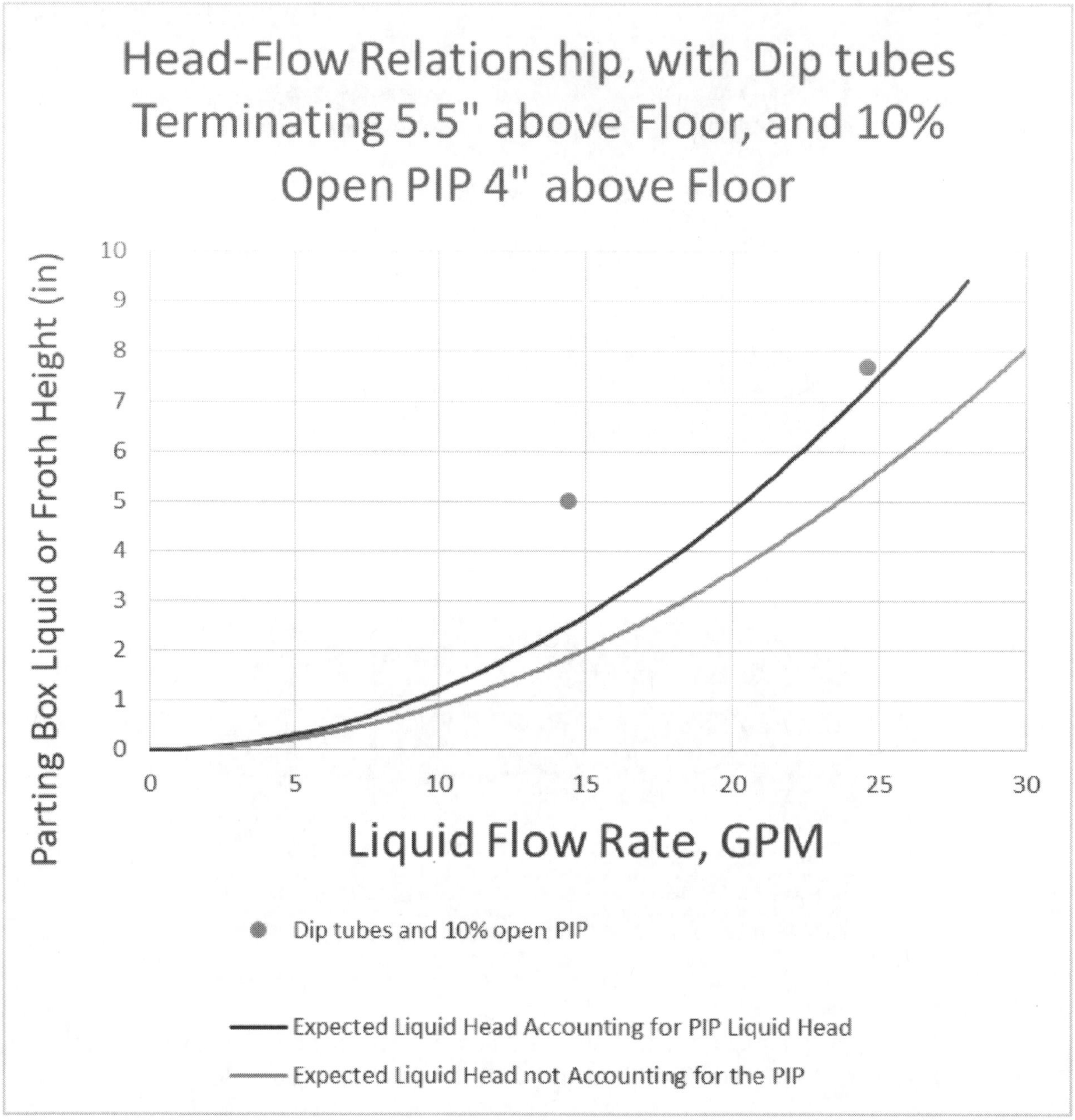

Figure 17. Head- flow relationship, with dip tubes 5.5" above the parting box floor, no tee at ends, and with 10% open PIP.

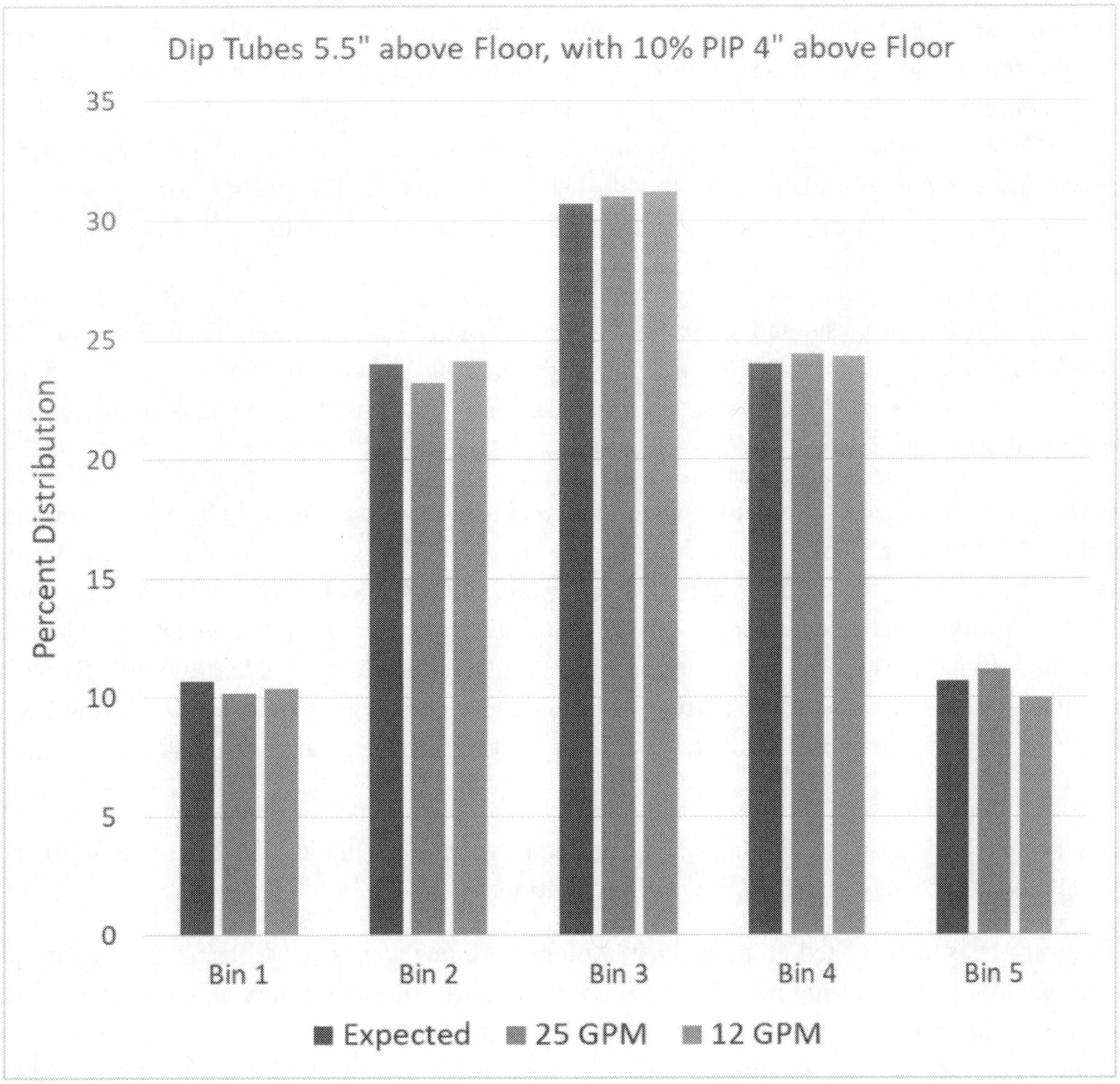

Figure 18. Liquid split among the bins compared to the desired split based of equal liquid flow rate through each perforation. Dip tubes terminating 5.5" above the parting box floor, 10% PIP mounted 4" above the floor.

Conclusions

We performed water tests on a scaled down model of a parting box that exists in a hydrocarbon-separation tower. In our previous work, we tested a feed sparger design scaled down from the hydrocarbon tower. The parting box, both in the actual tower and in the tests, had a design that broadly complied with published good practice guidelines.

Our previous tests[1] showed a highly frothed up liquid in the parting box at both high and low liquid rates. The frothing in the parting box was due to water jets issuing from the feed sparger pipe entraining air into the water upon impact. This generated excessive froth heights in the parting box, and at high rates, premature overflow of froth out of the parting box. The liquid split to the five clusters of holes at the floor of the parting box was good at the maximum rates but quickly deteriorated as the liquid rate was turned down. Using a Perforated Impingement Plate (PIP) with either 20% or 10% open areas, mounted either 2" or 4" above the parting box floor, greatly improved the split of liquid to the clusters of holes at the floor of the parting box. This split was very good to excellent both at the maximum rates and at turndown to about 50% of the maximum rates. Lower rates were not tested.

This work extended our tests into lower hole velocities out of sparger pipes. The lower velocities somewhat reduced aeration and froth heights in the parting box, but not by much. The split of liquid to the clusters of holes in the parting box was again good at the maximum rate but deteriorated upon turndown.

This work also extended our tests to a design in which dip tubes were used to bring the incoming liquid 1.75" above the floor of the parting box. This fully eliminated the frothing in all but the lowest rate, but again the liquid heights in the parting box were considerably higher than those calculated from the orifice equation. It may be that the horizontal momentum generated when the liquid jets hitting the floor of the parting box were deflected sideways made it more difficult for the liquid to descend through the parting box holes, causing larger liquid heads. At the lower rates, the dip pipes also appeared to cause air to be sucked into the sparger pipe via its vent holes.

Fitting the tubes with tees that directed the liquid towards the parting box walls was shown to have eliminated the excessive liquid height problem, gave good liquid distribution, and a head-flow relationship closely matching the orifice equation.

Our current tests also looked at a design in which dip tubes were used to bring the liquid 5.5" above the floor of the parting box. At the higher rates, with the tubes submerged in the parting box liquid, the liquid heights in the parting box were again considerably higher than those calculated from the orifice equation. In this case too, fitting the tubes with tees that directed the liquid towards the parting box walls gave good liquid distribution, and a head-flow relationship closely matching the orifice equation.

With the dip tubes bringing the liquid 5.5" above the floor of the parting box, adding a PIP with 10% open area with the tubes liquid-submerged, there was no frothing, no air gap under the PIP, and the liquid height in the parting box matched the height calculated from the orifice equation.

References

1. Kister, H. Z., G. Jacobs, and A. A. Kister, "Gaining Insights into Parting Box Hydraulics", in "Kister Distillation Symposium", Topical Conference Proceedings, p. 463, AIChE Spring Meeting, New Orleans, LA, March 31- April 4, 2019.
2. Norton Company, "Packed Tower Internals", Bulletin TA-80R, 1974 (available from KochGlitsch, Wichita, Kansas).
3. Fadel, T. M. "Selecting Packed-Column Auxiliaries", Chem. Eng., January 23, p.71, 1984.
4. Kister, H. Z. "Distillation Operation", McGraw-Hill, N.Y., 1990.
5. Senecal, V. E.,"Fluid Distribution in Process Equipment", Ind. Eng. Chem. 49(6), p.993, 1957.
6. Kister, H. Z., P. Mathias, D. E. Steinmeyer, W. R. Penney, V. Monical, and J. R. Fair, "Equipment for Distillation, Gas Absorption, Phase Dispersion, and Phase Separation", Section 14, in R. H. Perry and D. Green "Chemical Engineers' Handbook", 9h Ed., McGraw-Hill, N.Y., 2018.

"Non-Traditional Approach to Improve Random Packing Performance is Confirmed at Two Independent Research Centers."

By

J. Antonio Garcia*, Bob Lobaugh & Tom Wilson
AMACS Process Tower Internals
&
A. Frank Seibert
James R. Fair Process Science & Technology Center (PSTC)

Presented at the

2020 AIChE® Spring Meeting and Global Congress on Process Safety

DISTILLATION SYMPOSIUM 2020
Packing Technology – New Developments II
Paper 215b

August 19th, 2020
Houston, Texas U.S.A.

AMACS
PROCESS TOWER INTERNALS

"Non-Traditional Approach to Improve Random Packing Performance is Confirmed at Two Independent Research Centers."

A Vision 2020 update is upon us and despite the maturity of Distillation, equipment improvements are still being achieved and some of the key technical barriers for distillation described in the Vision 2020 Separations Roadmap are still present.

A non-traditional high-performance random packing (AMACS SuperBlend™) has been tested in the largest independent distillation research unit, Fractionation Research, Inc. (FRI).

The tests were performed under high pressure distillation with the iso-butane/normal butanes standard test mixture. These recent experimental results confirm the improved performance previously measured at the James R. Fair Process Science and Technology Center (PSTC) at The University of Texas at Austin, formerly known as the Separations Research Program (SRP)

This performance exhibited by AMACS SuperBlend™ provides an advanced and reliable alternative for the Chemical Process Industry (CPI) to debottleneck distillation and absorber columns limited by separation efficiency and capacity.

Experimental results and the non-traditional random packing design methodology will be presented.

"Non-Traditional Approach to Improve Random Packing Performance is Confirmed at Two Independent Research Centers."

J. Antonio Garcia* (drjagarcia@gmail.com), Bob Lobaugh and Tom Wilson
AMACS Process Tower Internals
14211 Industry Street. Houston, Texas 77053 USA

A. Frank Seibert (fseibert@mail.utexas.edu)
James R. Fair Process Science & Technology Center (PSTC)
EME Bldg. 133. 10100 Burnet Road. Austin, Texas 78758 USA

AMACS
PROCESS TOWER INTERNALS

Authors Information

Dr. J. Antonio Garcia*

Phone: +1 (832) 459-1994
Email: drjagarcia@gmail.com

Bob Lobaugh

Phone: +1 (281) 716-1881
Email: blobaugh@amacs.com

Tom Wilson

Phone: +1 (713) 332-0382
Email: twilson@amacs.com

AMACS Process Tower Internals

14211 Industry Street Houston, Texas 77053 USA

Dr. A. Frank Seibert

Phone: +1 (512) 471-7063
Email: fseibert@mail.utexas.edu

James R. Fair Process Science & Technology Center (PSTC)

EME Bldg. 133, 10100 Burnet Road

Austin, Texas 78758 U.S.A

AMACS
PROCESS TOWER INTERNALS

Presentation Outline

1. Vision 2020 – Distillation
2. Background
3. Traditional versus Non-traditional Approach
4. Independent Research Centers Experimental Results
5. Non-Traditional Design Methodology
6. Conclusions
7. References

AMACS
PROCESS TOWER INTERNALS

Vision2020

KEY TECHNICAL BARRIERS

PHYSICAL FUNDAMENTALS (H)	EQUIPMENT PERFORMANCE (M) (H=high priority, M=medium priority)	EDUCATION/INFORMATION TRANSFER (M)
Lack of accurate real stage efficiency calculation	Flow control on trays is inadequate	Companies and vendors will not share information
Lack of adequate mixing characterization	Liquid distribution in packed beds is inadequate	Industrial R&D funds are being reduced
Lack of vapor liquid imaging techniques	Scanning methods for tower operation troubleshoooting are inadequate	Universities and National labs are not emphasizing distillation research
"Bad actor" (foam) separation is inadequate	Lack of non-fouling distributors	Universities are reducing courses in distillation
Lack of ability to separate multi-phase flow traffic	Lack of active devices for phase disengagement	
A beeter understanding of mass transfer and multiphase flow is required	Lack of adequate means to prevent fouling in the towers	
A better process of the understanding of phase formation , mixing, interface area formation, and mass transfer is required	Lack of methods to handle multi-phase feeds	
Lack of adequate data on bubble formation	Equipment size-determining rate processes need to be re-evaluated	
	Hybrid column internals are required	
	Means of reducing stage spacing to less than 2 inches are required	

AMACS
PROCESS TOWER INTERNALS

Vision2020

KEY RESEARCH NEEDS

TIME FRAME	FUNDAMENTAL DATA (H=high priority, M=medium priority)	PROCESS SYSTEMS
ALL (ONGOING PROCESSES)	N/A	N/A
NEAR-TERM (0-3 YEARS)	Develop sensors for measuring basic distillation phenomena (H)	Develop column internals imaging equipment (M)
	Develop database of mass transfer and hydrodinamic data (H)	Develop better simulators (M)
		Develop database of existing packing/trays (M)
		Develop tools to predict composition from column temperature (M)
MID-TERM (3-10 YEARS)	Improve understanding of basic distillation phenomena (H) [Mixing, interfacial area, mass transfer, multi-phase flow]	Develop sensors for multi-phase flow (M)
		Develop understanding of factors affecting vapor/liquid distribution (M)
	Develop fundamental-based model for predicting mass transfer and hydrodynamics in complex "difficult" systems	Develop *in-situ* sampling methods (M)
		Develop better process gas chromatographs (M)
		Develop novel phase separation methods (M)
LONG-TERM (10+ YEARS)	N/A	N/A

AMACS
PROCESS TOWER INTERNALS

Background

- In recent years 4th generation random packings are becoming the standard for new column designs.
- There are several vendors offerings with "newer" packing geometries that enhance mass transfer and hydraulic performance compared to conventional packings (3rd generation).
- Available 4th generation packings have been developed by US or European companies.
- A well-established 3rd generation random packing widely used in the chemical process industry is now been replaced by latest generation packings.
- This well-established 3rd generation packing is the basis for a non-traditional approach to enhance mass transfer and hydraulic performance of random packings.

8

Distillation Symposium 2020 : Packing Technology – New Developments II
215b. J. A. Garcia, B. Lobaugh, T. Wilson and A. F. Seibert

AMACS
PROCESS TOWER INTERNALS

Traditional Random Packing Evolution

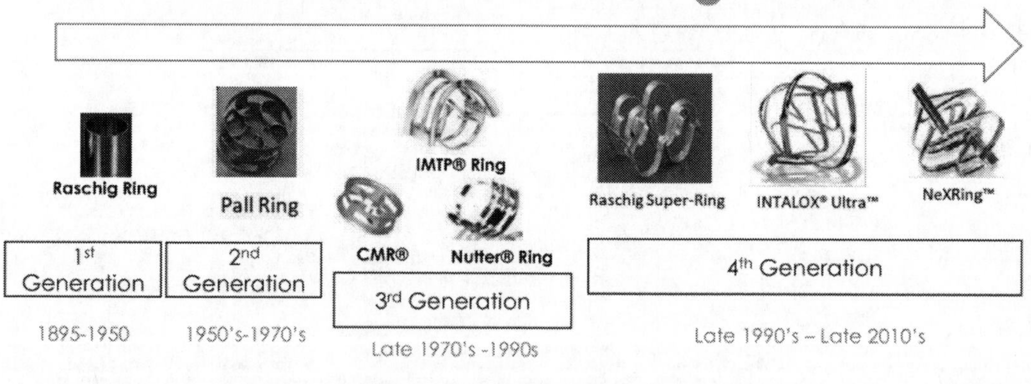

Distillation Symposium 2020 : Packing Technology – New Developments II
215b. J. A. Garcia, B. Lobaugh, T. Wilson and A. F. Seibert

9

AMACS
PROCESS TOWER INTERNALS

Traditional vs. Non-Traditional Approach

Traditional

- Different packing geometry determine vapor-liquid interaction
- Packings are same size and shape
- This specific geometry ("shape") is characterized
- Lower pressure drop per unit length and stable mass transfer efficiency

Non-traditional

- An optimized blend of different sizes of random packing determine vapor-liquid interaction
- Packings are same shape
- The optimized blend is characterized
- Capacity near to the larger packing size at an improved mass transfer efficiency

AMACS
PROCESS TOWER INTERNALS

4th Generation Random Packings

Raschig Super-Ring

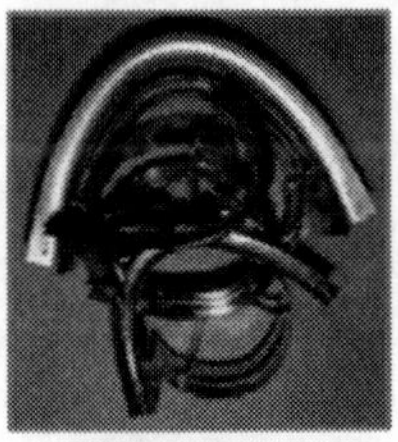

SUPERBLEND 2-PAC™

INTALOX® Ultra™

NeXRing™

AMACS
PROCESS TOWER INTERNALS

Non-traditional Random Packing

AMACS SuperBlend™

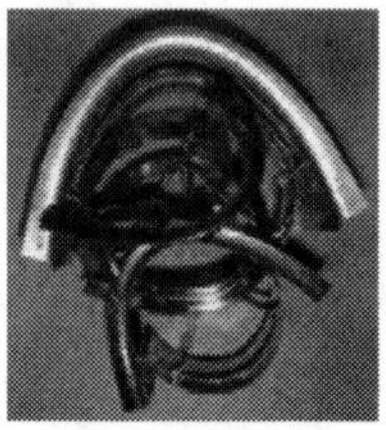

1st Generation
2-Pac™

2nd Generation
ASB™

AMACS
PROCESS TOWER INTERNALS

Distillation Units General Details

PSTC

- Diameter: 18-in (0.457 m)
- Total Reflux distillation and [L/V] <>1
- Operating pressures:
    - Vacuum up to 60 psia (4.14 bara)
- Systems:
    - Cyclohexane/n-heptane
    - Xylenes
    - Ethylene glycol/water
- Devices: Trays and packings
- AMACS SuperBlend™ tests
    - 2004, 2006, 2017-2020

FRI

- Diameter: 4-ft and 8-ft (1.2 & 2.4 m)
- Total reflux distillation and [L/V] <>1
- Operating pressures:
    - Vacuum up to 500 psia (34.5 bara)
- Systems
    - Butanes
    - Cyclohexane/n-heptane
    - Xylenes, and others
- Devices: Trays & packings
- AMACS SuperBlend™ test
    - 2019

AMACS
PROCESS TOWER INTERNALS

Independent Research Centers

James R. Fair PSTC Pilot Plant

FRI Unit

AMACS
PROCESS TOWER INTERNALS

AMACS SuperBlend™ Testing at PSTC

SuperBlend Random Packing Studies

PSTC Standard aqueous solution/air Testing at atmospheric conditions

- Capacity
- Pressure drop
- Liquid holdup, and
- Effective mass transfer area

Estimated Packing Surface Area (a_p)		
Packing Run	(m^2/m^3)	(ft^2/ft^3)
SB0	116	35
SB1	145	46
SB2	76	23
SB3	144	44
SB4	133	40
SB5	133	40
SB6	116	35

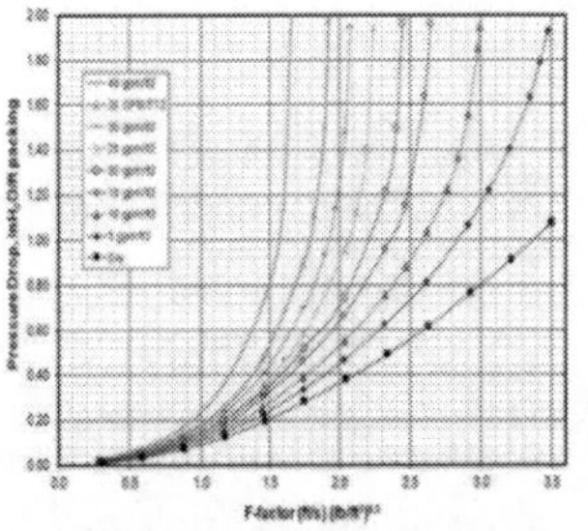

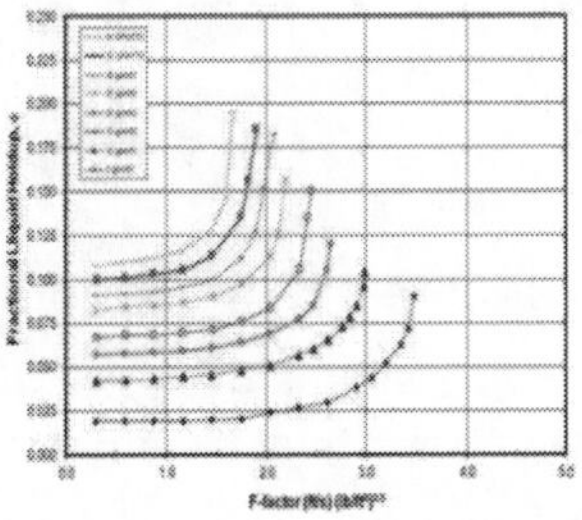

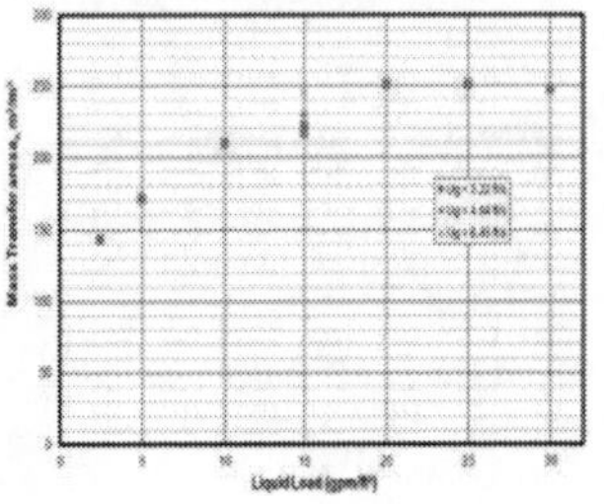

AMACS
PROCESS TOWER INTERNALS

PSTC Tests Performance Comparison

A. 2" Pall Rings
B. 1" Pall Rings
C. SuperBlend™ ASB™ 50

- Standard testing protocol
- Similar internals and set-up
- Cyclohexane/n-heptane
- Total Reflux
- 2.4 & 60 psia

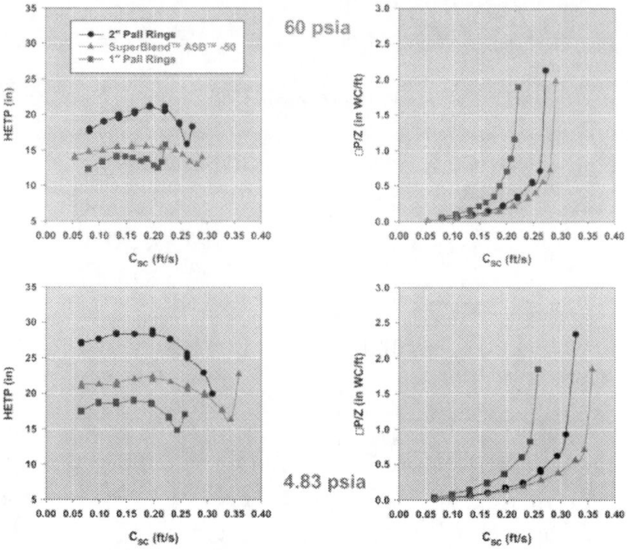

Distillation Symposium 2020 : Packing Technology – New Developments II
215b. J. A. Garcia, B. Lobaugh, T. Wilson and A. F. Seibert

16

AMACS
PROCESS TOWER INTERNALS

AMACS SuperBlend™ Random Packing Performance Comparison

A. SuperBlend™ ASB™ 50
B. SuperBlend™ 2-Pac™ #4050

- Standard testing protocol
- Similar internals and set-up
- Cyclohexane/n-heptane
- Total Reflux
- 4.83 & 60 psia

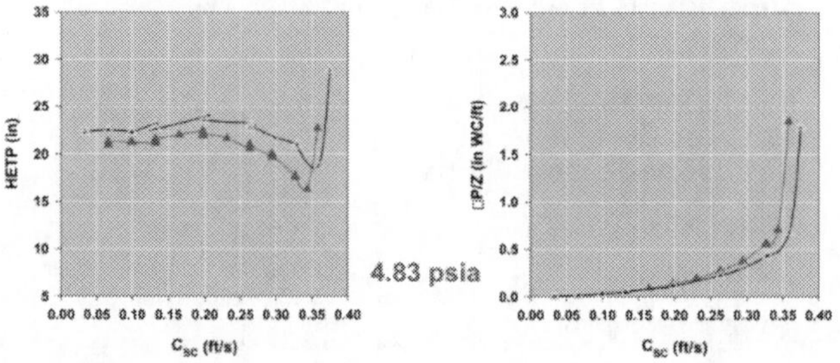

Distillation Symposium 2020 : Packing Technology – New Developments II
215b. J. A. Garcia, B. Lobaugh, T. Wilson and A. F. Seibert

17

AMACS
PROCESS TOWER INTERNALS

PSTC Experimental Results

- Hydraulic and effective surface area measurements confirmed the improved performance of packing blends.
- Total reflux distillation confirmed the efficiency improvements for the blended packing with 10-15% efficiency improvements with a small penalty on packing capacity.
- Benchmarking 2nd and 3rd generation packings confirmed efficiency improvements.
- 2nd Generation SuperBlend random packing exhibited a 10-15% improved efficiency at capacity near the 1st generation or larger saddle ring size.
- Rukovena and Koshy Method assisted in planning packing blend a/w and distillation testing.

AMACS
PROCESS TOWER INTERNALS

FRI Test (2019) Installation Activities

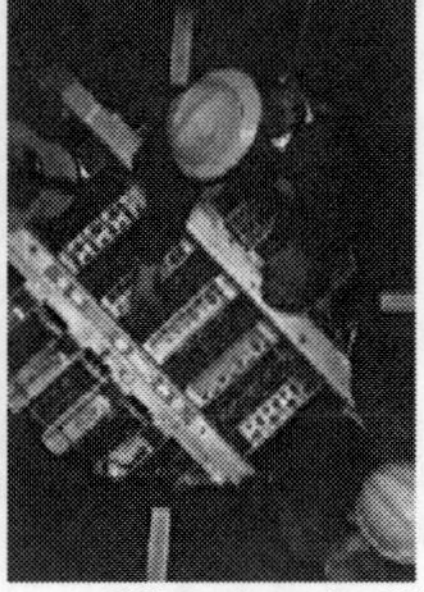

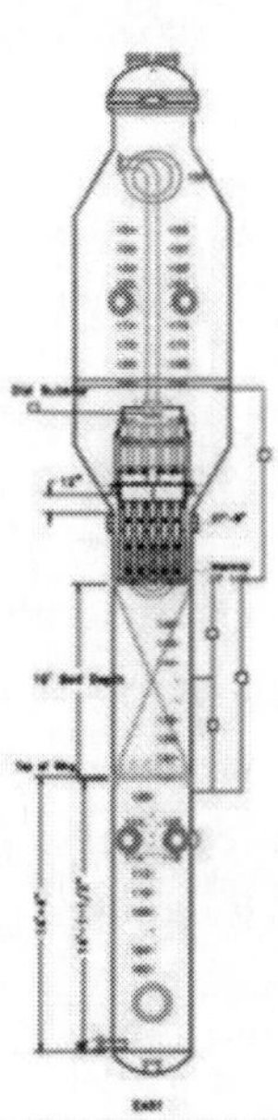

AMACS
PROCESS TOWER INTERNALS

FRI TEST

- AMACS SuperBlend™ ASB™
  - Nominal size: 50 mm
- Column: Low Pressure (LP)
- Diameter: 4-ft (1.22 m)
- Bed depth: 10-ft (3.05 m)
- System: Butanes (i-C4/n-C4)
- Pressure: 165 psia (11.4 bara)
- Standard testing protocol
- Internals by AMACS

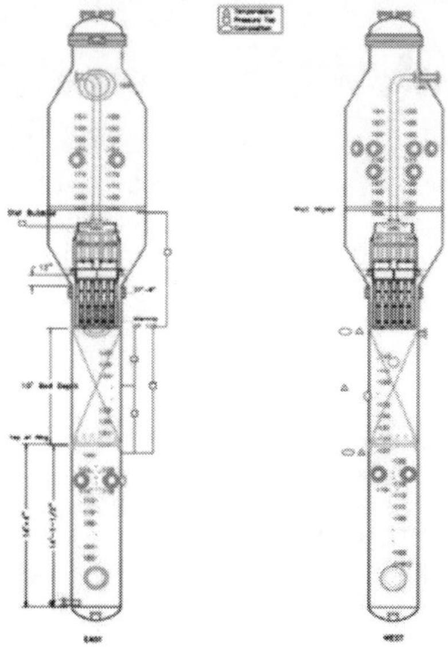

Distillation Symposium 2020 : Packing Technology – New Developments II
215b. J. A. Garcia, B. Lobaugh, T. Wilson and A. F. Seibert

20

AMACS
PROCESS TOWER INTERNALS

FRI Test Operations Diagram & Efficiency

- SuperBlend™ ASB™ 50
- FRI standard testing protocol
- System: Butanes (i-C4/n-C4)
- Pressure: 165 psia (11.4 bara)
- Total Reflux [(L/V) = 1]
- Hydraulic [(L/V) <> 1]

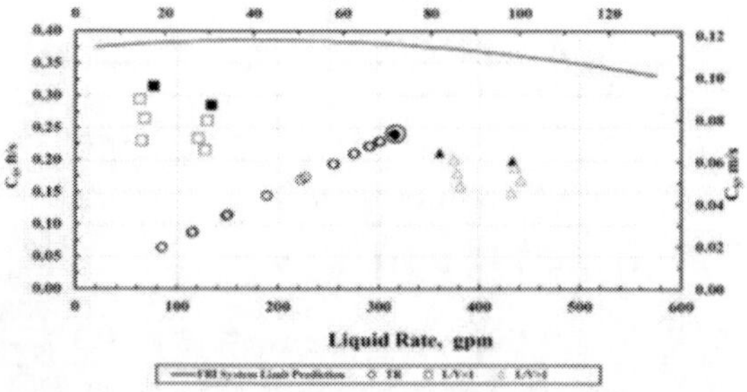

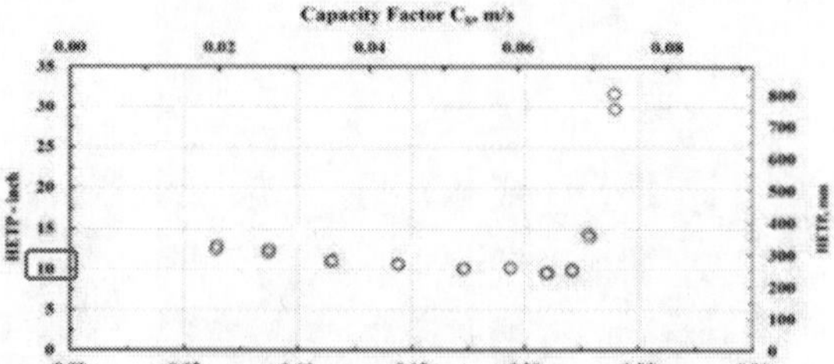

Distillation Symposium 2020 : Packing Technology – New Developments II
215b. J. A. Garcia, B. Lobaugh, T. Wilson and A. F. Seibert

21

AMACS
PROCESS TOWER INTERNALS

AMACS SuperBlend™ ASB™ 50 Performance at FRI & PSTC

Total reflux

PSTC

- **Cyclohexane/n-Heptane**
- **50-60 psia (3.45 & 4.14 bara)**

FRI

- **Iso-Butane/n-Butane**
- **165 psia (11.4 bara)**

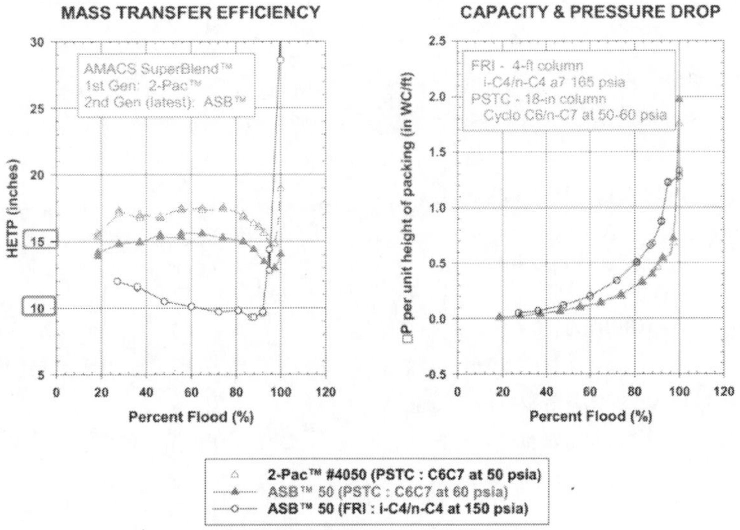

AMACS
PROCESS TOWER INTERNALS

FRI Experimental Results

- AMACS SuperBlend™ ASB™ 50 random packing efficiency ranged from 9.8 to 12.1 inches (249 to 307 mm) with the C4's system at 165 psia (11.4 bara).
- FRI experimental results with butanes re-confirmed mass transfer improvements and hydraulic performance previously measured at PSTC unit with cyclohexane/n-heptane.
- Experimental results were released to FRI members under Topical Report 212.

AMACS
PROCESS TOWER INTERNALS

AMACS SuperBlend™ Capacity Chart Nominal size: 50

FRI & PSTC Experimental Results

- **Iso-Butane/n-Butane**
  - **165 psia (11.4 bara)**
- **Cyclohexane/n-heptane**
  - **2.4-60 psia (0.17-4.14 bara)**

SuperBlend™ Capacity Chart - Nominal size 50

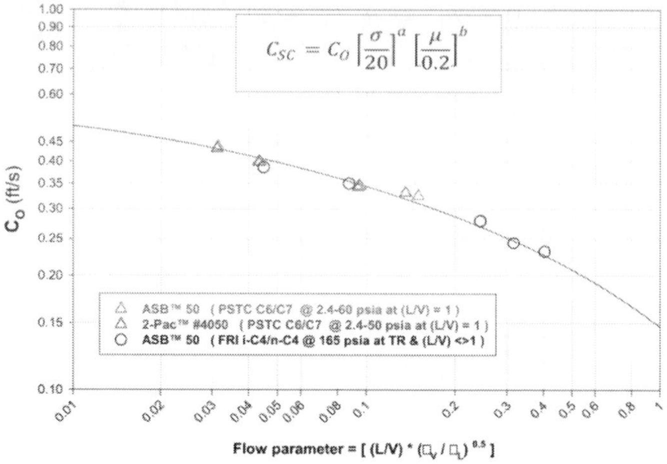

AMACS
PROCESS TOWER INTERNALS

Conclusions

- A blend of different size random packings exhibit improved mass transfer performance with capacity similar to the larger size packing.
- Recent AMACS SuperBlend™ ASB™ **50 random packing** tests at Fractionation Research, Inc. (FRI) confirmed experimental results from the James R. Fair Process Science & Technology Center.
- Available sizing methods for saddle ring packings are amenable to packing blends of different size packings.
- Rukovena and Koshy sizing method used for 3rd generation random packing (saddle rings) can be easily applied when designing packed columns using blended saddle ring packings.
- AMACS will continue sharing experimental results of the non-traditional approach to enhance the mass transfer performance of random packing blends.

AMACS
PROCESS TOWER INTERNALS

Acknowledgements

- James R. Fair Process Technology Center
  - Technical Staff
  - Dr. Bruce Eldridge and Dr. Frank Seibert
- Fractionation Research, Inc.
  - Technical Staff
  - Dr. Ken McCarley , Dr. Tony Cai and Mark Cavett

AMACS
PROCESS TOWER INTERNALS

References

1. VISION 2020 1998 Separations Roadmap. Published by AIChE and US Department of Energy Office of Industrial Technologies, DOE/OIT (1998)
2. Michael Schultes. "Raschig Super-Ring. A New Fourth Generation Packing Offers New Advantages." Trans IChemE. Vol 81, Part A. January 2003.
3. Frank Rukovena and T. Daniel Koshy. "Packed Distillation Tower Hydraulic Design and Mechanical Considerations." Ind. Eng. Chem. Res. 1993, 32, 2400-2407.
4. A. Frank Seibert. "Distillation and Air/Water Performance Characteristics of the AMACS SuperBlend™ Random Packings." James R. Fair Process Science and Technology Center (June 2019).
5. J. Antonio Garcia, Bob Lobaugh , Tom Wilson and A. Frank Seibert. "Improving Mass Transfer and Hydraulic Performance of Random Packings by a Non-traditional Approach." 2019 AIChE Spring National Meeting. Distillation Symposium. "Exciting New Developments in Packings Technology." Session 109, Paper 4 f. New Orleans, Louisiana USA (April 2nd, 2019).
6. J. Antonio Garcia, Bob Lobaugh , Tom Wilson and A. Frank Seibert . "A Non-traditional Approach that Improves Random Packing Performance." 11th AIChE Southwest Process Technology Conference. Distillation & Separation. October 1st, 2019. Sugarland, Texas U.S.A.
7. T. Cai and M. Cavett. "Test of AMACS SuperBlend™ Random Packing." Fractionation Research, Inc. Topical Report 212. August 3rd, 2020.

Thank you!

Q&A

592114 Process and adsorbent for separating ethanol and associated oxygenates from a biofermentation system

Abstract:

Abstract for poster presentation or talk:

Abstract Title: Process and adsorbent for separating ethanol and associated oxygenates from a biofermentation system

Presenter Information: Deepak Sharma; Sr. Process Engineer, Engineering and Technology, Bayer AG

Patent number: US8658845B2

Patent Authors:

Deepak Sharma - Sr. Process Engineer, Bayer Ag, St. Louis; (Presenting Author)

Anil Oroskar – CTO, Orochem Technologies, Napervielle, IL;

David W. House – Scientist, Orochem Technologies, Napervielle, IL;

Alice M. Havill – Scientist, Lanzatech, New Zealand

The principles of process intensification if used on existing processes can lead to development and redesign of existing processes to become more sustainable. This presentation will discuss a process and an adsorbent for the separation of ethanol associated oxygenates from a dilute mixture of ethanol and associated oxygenates in water in the presence of organic compounds derived from a biofermentation process. After pretreatment, the separation is carried out in a simulated moving bed adsorption system employing a stationary phase adsorbent comprising fluorinated carbon or modified C18 silica gel selective for the adsorption of ethanol and associated oxygenates, such as 2,3-butanediol, with a mobile phase desorbent selected from the group consisting of methanol, ethanol, propanol, and methyl tertiary butyl ether. The process is useful for removing water from dilute aqueous mixtures of organic compounds comprising ethanol in dilute concentration in water and produced by fermentation, biomass extraction, biocatalytic, and enzymatic processes which are not economically recoverable by conventional distillation methods.

Keywords: Separation; Adsorption; SMB; Purification

Patent Assignee: Orochem Technologies LLC

592111 Pump safety, the undiscussed topic in chemical industry leading to several fatal incidents

Abstract:

Abstract Title: Pump safety, the undiscussed topic in chemical industry leading to several fatal incidents

Presenter Information: Deepak Sharma; Sr. Process Engineer, Engineering and Technology, Bayer AG

Pumps are one of the equipment's in the process industry that require a thorough review during HAZOP studies. The constant source of energy given to the pump, if not converted to kinetic energy to flow the fluid, can lead to heating of the pumped fluid. If the fluid keeps accumulating the energy provided by the motor, in a blocked-in scenario, it can lead to very high temperatures eventually leading to loss of containment of the pump and the pump can get damaged. There have been instances in the industry where fragments of damaged pumps have flown several feet away from the point of loss of containment leading to injuries. The poster will outline a scenario where a pump is run blocked in on suction and discharge for water and a chemical which has a low thermal decomposition temperature. Arc data is used to evaluate the onset temperature of thermal decomposition. The poster will outline safeguards that can be employed to avoid catastrophic loss of containment of the pump. Outlined below are safeguards used to avoid loss of containment of a pump in case of a blocked-in scenario:

- Install PSV on suction of pump
- Employ a low amp/power interlock on the pump to turn off the pump in case of blocked in situation
- Have a high-pressure interlock on the discharge line of the pump to direct flow back to the tank in case of pump deadhead
- Install an Interlock to turn off the pump in case the feed forward valves (automatic valves) close inadvertently

Pump deadhead scenarios are not only catastrophic when in flammable service there are several salts that have low thermal decomposition temperatures. Pumps in water service left blocked in could also blow catastrophically leading to reportable incidents. I feel this is an important topic to bring to the attention of students and professionals in the chemical industry. Poster will include pictures of actual installed pumps along with required safeguards in the field along with supporting P&ID depictions.

589440 New Kind of Linear Nanocomposite Applied in Improving Rheological Behavior of Crude Oil and Its Mechanism Studied through Molecular Simulation Method

Abstract:

In this work, a new kind of linear nanocomposite produced by carbon nanotube(CNT) as pour point depressant (C2-100) was developed by blending with MCNT nanoparticle and copolymer named Ethylene-vinyl acetate (EVA). Its effect of reducing yield stress and modifying the morphology of wax agglomerate were also evaluated. Besides, the synergetic effect of C2-100 nanocomposite with the magnetic field was also conducted. The mechanism of modifying rheological behavior of waxy crude oil was studied through molecular simulation method, which aims at understanding the interaction between this nanocomposite and wax molecule. The results indicated that C2-100 nanocomposite could reduce the yield stress of waxy crude oil more effectively than the copolymer (EVA) even at low concentration, as well as changing the morphology of wax crystals and agglomerate. Moreover, its effect of reducing yield stress becomes more obvious when the alternate magnetic field was applied. In addition, the results of molecular simulation also presented that the effect of improving rheological behavior of C2-100 nanocomposite is related to the adsorption behavior of wax molecules on the surface of MCNT nanoparticle, which is adsorbed with EVA molecules. In addition, the adsorption behavior of different fragments of EVA molecules can also influence the adsorption behavior of wax molecules on the surface of EVA-attached MCNT nanoparticle, which may influence the effect of C2-100 to improve the rheological behavior of waxy crude oil.

589683 Solar Energy Conversion and Noise Characterization Sensing in Building Integrated Solar Photovoltaic Devices with Ventilation

Abstract:

A study on solar energy conversion and noise characterization sensing in solar photovoltaic devices with ventilation is conducted. A building integrated parallel plate solar photovoltaic (PV) device was built with a pair of PV modules, a ventilated air cavity and an insulating back panel of plywood board filled with polystyrene installed in an outdoor test-room. Photovoltaic amplification was attained with power output from a potentiometer through rotation of its circular knob. A parallel plate PV device was studied for its
electrical parameters as Resistance-Capacitance (RC) electrical analog circuit. The effect of inductive and capacitive heating losses was considered in evaluating electrical characteristics of a PV device exposed to solar radiation. The characterization of noise interference, due to power difference of two intensities in a wave on PV Device is presented. Noise characterization of a wave is obtained depending on type of wave. The difference of two power intensities is due to transmission of light, sound, heat, electricity, fluid, and fire into a parallel plate PV device. Based on analysis of a solar PV device, a definition with measurement expression and unit for an electrical noise were coined. Standard definitions of noise sources, their measurement equations, their units and their origins under limiting reference conditions were also devised. A review for sensing noise of physical agents is conducted for sensing noise of physical agents (sun, light, sound, heat, fluid, electricity and fire). With integrated environmental control, these environmental parameters are to be characterized through various environmental monitoring sensors.

589945 Structural Characterization and Model Representation of Vitrinite-Rich India Coal By Using Advanced Analytical Techniques

Abstract:

The detailed investigation of the chemical structure of Karharbari coal (vitritine rich subbituminous coal from India) by utilizing advanced instrumental analytical techniques such as solid state ^{13}C nuclear magnetic resonance (^{13}C-NMR) spectroscopy, Fourier transform infrared (FT-IR) spectroscopy, X-ray diffraction (XRD), Flash-pyrolysis gas chromatography mass spectroscopy (Py-GC/MS), Laser desorption ionization time-of-flight mass spectroscopy (LD-TOF-MS), X-ray photoelectron spectroscopy (XPS), and high-resolution transmission electron microscopy (HRTEM). The Karharbari coal was an amorphous loosely organized lower rank coal with aromaticity 0.79. This was indicated by XRD results and also confirmed with ^{13}C-NMR results (aromaticity =0.84) and FTIR results (aromaticity = 0.82). The average molecular weight of the aromatic cluster is 507 amu by NMR result which is closer to the result obtained by HRTEM (MW = 530.46 amu). The presence of different heteroatoms like carbon, oxygen, nitrogen and sulphur with their functionalities, were determined by using a XPS technique. Different carbon/oxygen forms present in the Karharbari coal are (C-C) and (C-H) 68.53%, (C-O) 23.46%, (COO^-) 1.96%, (C=O) 6.05% and different nitrogen forms such as pyridine, pyrrolic, quaternary, oxidized and their distribution (mol %) are 19.36%, 45.68%, 31.25%, 3.71 respectively, and different sulphur forms were present i.e. corresponding to thiophenes, sulphones, sulphuric acid and sulphates with the molar content of 16.47%, 41.61%, 21.37%, 20.55% respectively. The construction of the individual molecules will be discussed in detail in the rest of this paper. The individual molecules were geometry-optimized and Materials Studio Amorphous Builder module was used to construct a three dimensional associated structure.

589479 Life Management of High Energy Piping with Highlights from ASME B31.1

Abstract:

High energy piping (HEP) systems – or as defined by the ASME B31.1 Power Piping Code, Covered Piping Systems (CPS) – are critical assets with respect to production facilities. A failure can have catastrophic consequences from a safety perspective and in terms of equipment loss and downtime. In fact, loss of high-pressure steam due to a failure in a main run of steam supply piping could shut an entire production unit down. Because of this and the requirements of the Code, HEP maintenance and inspection programs are established to ensure that the integrity of the systems is maintained in a safe and reliable condition throughout their lifecycle. There is no one-size-fits-all for the life management of these critical assets because there is a great variety in age, design, materials, and owner objectives. Life management of these assets is beneficial in that it provides a structured, systematic approach to creating and implementing a program including evaluating the current program (whether formal or informal), what the objectives of the program are (short and long term), what equipment is in place, and its history.

A HEP program put in place for overall life management is not synonymous with an inspection program, although inspections are an important part of an overall program. Inspections should be used in conjunction with analytical tools, real time monitoring, and laboratory examinations. The overall approach is life management – not inspection management – of the equipment.

This paper describes a HEP life management approach which includes identifying locations of interest, risk-ranking those locations, implementing a plan for determining the health of the highest risk locations, and taking actions based on the system health that has been determined. This paper also presents and discusses some of the requirements of the B31.1 code, which at the highest level states that each operating company shall develop operation and maintenance procedures for piping systems deemed necessary to ensure safe facility operations based on the provisions of the code, relevant industry experience, the operating company's experience (and knowledge of its facility), and the conditions under which the piping systems are operated.

590131 Optimization of C3MR LNG Liquefaction Process Using a Parallelized Simulator

Abstract:

Controlling the cost of energy-intensive processes such as C3MR (propane/mixed refrigerant) LNG liquefaction is a prime opportunity for optimization. Our work combines different C3MR models to develop an optimization that uses multiple independent variables, relaxed constraints, and flowsheet alterations to achieve an optimal operating cost. This work uses a rigorous model of the process in a process simulator rather than a reduced model of the process response. Use of a first-principles process simulator increases the accuracy of the model and therefore allows a more realistic optimization, at the cost of increased computing time. Parallelization (using multiple CPU cores) in the process simulator greatly increases the speed of the optimization calculations. CHEMCAD's optimization tools and multi-core technology make it possible to easily define the optimizations, run the optimizations efficiently, and generate global optima with a high degree of confidence.

590196 Modifying Fischer-Tropsch Reaction Sequence with Addition of Zeolite to Iron-Based Activated-Carbon Supported Catalyst

Abstract:

In addition to previously presented scientific work, modified reaction sequence of Fischer-Tropsch (FT) process was studied by addition of Zeolite ZSM-5 to an iron-based activated-carbon-supported catalyst. Main reactions involved due to FT catalyst are Fischer Tropsch Synthesis (FTS) and Water Gas Shift (WGS) reactions. Main reactions due to ZSM-5 are cracking, oligomerization, isomerization, cyclization, aromatization, conversion of alcohols to hydrocarbons, and coke formation. The waxy and linear structure FT products are converted to gasoline range, lighter hydrocarbons, iso-paraffins, and aromatics using ZSM-5. The catalyst ratio of 1:2, FT catalyst: ZSM-5, is used at 300 PSIG and 300°F. Separate-bed configuration, i.e. FT and ZSM-5 are separated in a reactor, arrangement is used for the reaction sequence analysis. Alcohols with carbon-chain from 3-5 produced by FTS are converted into hydrocarbons over ZSM-5. Waxy hydrocarbons with carbon-chain greater than 23 produced by FTS are cracked and isomerized to small n-paraffins and iso-paraffins using ZSM-5. The hydrocarbon with carbon chain from 7-10 produced by ZSM-5 cracking reactions are converted to aromatic products over ZSM-5. Produced gasoline to kerosene range n-paraffins are further cracked into smaller chain (1-5) iso- and n-paraffins. Some of the iso-paraffins are oligomerized and aromatized over ZSM-5 to form more aromatics. Further, multiple aromatic rings combine to undergo cyclization and dehydrogenation reactions over ZSM-5 to form coke. Water produced as a byproduct of FTS is combined with some of the carbon monoxide, which is the WGS, to form carbon dioxide and hydrogen.

590427 Lateral Retention Force Required for Sliding Saline Droplets over Silianized Silicon Surface

Abstract:

The lateral retention force required to slide a liquid drop on a surface is expected to increase with the time that the drop rests on the surface prior to sliding till it reaches a plateau. The molecular reorientation of the molecules at the periphery of the solid surface subjected to the unresolved vertical component of surface tension force causes this phenomenon. In this system, the surface tension of deionized water is increased by adding NaCl and is expected that the lateral retention force is increased proportionally with concentration. The retention force increased with increase in concentration reaching a maximum only to drop to a minimum with concentrations closer to saturation. That is, the lowest sliding force is required for both lower and higher concentrations of NaCl. This sudden decrease in observed lateral force is not a function of surface tension and the surface on which the drop slides. This phenomenon is associated with adhesion of salt to the surface and is supported by the drop sliding motion and depinning of the liquid from the solid.

590433 Study of Droplet Motion across Polymeric Substrates By Centrifugal Adhesion Balance (CAB)

Abstract:

On a given surface, the contour of the drop changes accordingly based on factors of the force applied, chemical structure of the fluid, drop size, and surface topography. With increasing force that is exerted on the drop, we determine the comprehensive trend for drop motion with respect to the controlled factors mentioned above. Drop motion is categorized into four phases and the point at where the initial displacement of receding line is observed is defined as the onset of motion. In this study, the onset of motion is affected by the factors mentioned above and aids in the determination for another value correlation known as Gs, known as Interfacial Modulus. The resistance to interaction towards the droplet as observed is quantified with respect to change in surface topography. This study therefore characterized drop motion and specifically onset motion based on controlled variables (factors). Work of adhesion is also quantified across different substrates to characterize the detachment of liquid from its substrate. All these parameters such as Gs, onset of motion, contact angles were analyzed with different polymeric substrates, drop sizes, and various fluids used.

590665 Population Balance Modeling and Spatially-Oriented Control for Crystal Size and Shape in Crystallization Processes

Abstract:

Crystallization is an important separation and purification process with relevant applications in the food, pharmaceutical, fine chemical and fertilizing industries. In these processes, modeling and process control for crystallization are focused on crystal size distribution (CSD) and shape of the crystals in order to make it possible to achieve the desired quality and purity of the product of interest. The control of crystal size and shape is of considerable importance for industrial crystallization processes, as crystal morphology has great influence on downstream steps and also from the point of view of product specification criteria. The consideration of crystal shape in modeling of crystallization processes is also important due to its impact on product effectiveness, such as the bioavailability and tablet stability in pharmaceuticals. Control of the crystallization process is can be facilitated with online measurements, which provide information to be used in the modeling of the process. However, if online measurements are not available, the strongly nonlinear dynamics of these processes makes difficult to establish relations between some inputs (e.g. temperature) and the final desirable characteristics of the crystal product. In this context, image analysis has been shown to be a suitable technique to monitoring the process [1].

In our present work, a population balance model-based scheme for controlling the size and shape of crystals is presented. The modeling was performed using a two-dimensional population balance (2D-PBM) in terms of the two characteristic dimensions of the crystals, describing the evolution of crystal size and shape distributions along the batch time. To identify the nucleation, growth and dissolution kinetics parameters using the 2D-PBM approach along the length and width directions, QICPIC-LIXELL's high speed dynamic image analysis equipment was used for the online measurement of the CSD, shape parameters and aspect ratio distribution of crystals during the experiments. In this way, the developed model allowed for the prediction of the crystal size and shape distributions during growth-dissolution cycles.

The developed framework was employed to obtain optimal temperature policies for controlling the crystal size and shape distributions that meet target criteria for size, morphology and productivity using the dynamic image analysis along the runs. Based on the concept of using measurements along batch crystallization processes as a trajectory in a phase space [2], a trajectory-endpoint control problem is presented with an oriented control scheme that is applied to the shape of the crystals. The scheme was used to control the batch cooling crystallization of potassium dihydrogen phosphate. The reported experimental data demonstrate the application of the control policies for temperature to produce crystals of the desired average shape and size in different batch times. Thus, using the first principle modeling coupled to the developed trajectory tracking control, the proposed framework has the potential to be applied to industrial crystallization processes for shape control and to be extended to polymorph control investigations.

References:

[1] Eisenschmidt, H.; Bajcinca, N.; Sundmacher, K. (2016) Optimal Control of Crystal Shapes in Batch Crystallization Experiments by Growth-Dissolution Cycles. *Crystal Growth & Design*, v. 16, p. 3297-2803.

[2] Griffin, D. J., Grover, M. A., Kawajiri, Y., Rousseau, R. W. (2015) Mass-count plots for crystal size control. *Chemical Engineering Science*, v. 137, p. 338-351.

590450 Auditing the Process Hazard Analysis - Key Characteristics to Scrutinize in a Good-Quality HAZOP Study and Supporting Effective HAZOP Study Updates

Abstract:

Although Process Safety Management (PSM), requiring good-quality Process Hazard Analyses, was promulgated in the United States in 1992, and HAZOP Study techniques date back to the mid-1970's, the HAZOP Study technique is still not uniformly applied in a high quality fashion. The objective of this paper is to provide experience-based tips on Process Hazard Analysis auditing, and some key characteristics that the auditor can scrutinize to assure quality implementation. Analogously, a similar set of characteristics that can facilitate future HAZOP Study updates can also contribute to a quality HAZOP Study that supports the spirit of Safety Management System objectives for the identification of hazards potentially impacting plant personnel, the community, and the environment. This paper will also review tips for structuring the HAZOP Study to facilitate later updates and revalidation.

590342 Development of High Performance Organic-Based Cathodes

Abstract:

For lithium-based batteries, structure/property relations in terms of cell voltage are desired for new organic-based cathode active materials. In this initial pursuit of structure/property relations, ab-initio calculations were performed on a few organic-based molecules to show the effect on the theoretical cell voltage by inclusion of nitrogen and/or oxygen. Preliminary test results are presented on a newly developed molecule and methods to potentially improve on the cell performance are given.

590705 Electrochemical Kinetics of Degradation of Graphite Anodes in Lithium Ion Batteries

Abstract:

Growth of the solid electrolyte interphase (SEI) is a major driver of capacity fade in LIBs. Despite its importance, the fundamental mechanisms remain unclear, primarily because of the complicated reaction pathways involved[1–3]. SEI growth can be both electrochemical and chemical in nature[4], and thus, it is a strong function of the potential and degree of lithiation of the electrode. We model the early-stage and long-term growth of SEI by accurately capturing the potential dependence of its formation kinetics as well as long term rate limiting steps, and validating it against the world's largest open source battery cycling data, generated in-house[5]. This is done using the Multiphase Porous Electrode Theory (MPET) framework[6] on graphite (phase separating) and carbon black (non phase separating) particles.

Lithium plating is another key degradation phenomenon that has been elusive, and it becomes important while trying to fast-charge batteries, i.e., 0% - 80% state-of-charge in 30 mins. We show that lithium plating is a key function of electrode morphology, phase-separation dynamics and potential. Phase-separation in graphite is modeled in the electrode using the Cahn-Hilliard Reaction framework described by Bazant[7]. We understand the electrochemistry of the onset of lithium plating with in-situ measurements connected to real time cell potential in a phase-separating electrode for the first time[8].

Results indicate that the peak SEI-forming currents are higher for higher driving currents. Also, we find that SEI only grows during electrode lithiation, i.e. the battery only degrades while being charged. We also find that onset of lithium plating is correctly captured only when phase separation in active material is accounted for. Further, the onset of plating is delayed on electrodes with a thick SEI layer – understanding SEI/plating coupling is integral to predicting fast charging manufacturing protocols for LIBs. This work holds promise for the predictive design of procedures[9] for manufacture and formation of LIBs.

[1] Cohen, Y. S.; Cohen, Y.; Aurbach, D. Micromorphological Studies of Lithium Electrodes in Alkyl Carbonate Solutions Using in Situ Atomic Force Microscopy. J. Phys. Chem. B 2000, 104, 12282–12291. https://doi.org/10.1021/jp002526b.

[2] Horstmann, B.; Single, F.; Latz, A. Review on Multi-Scale Models of Solid-Electrolyte Interphase Formation, Current Opinion in Electrochemistry 13, 62-69 2019.

[3] Nie, M.; Abraham, D. P.; Seo, D. M.; Chen, Y.; Bose, A.; Lucht, B. L. Role of Solution Structure in Solid Electrolyte Interphase Formation on Graphite with LiPF6 in Propylene Carbonate. J. Phys. Chem. C 2013, 117 (48), 25381–25389, https://doi.org/10.1021/jp409765w.

[4] Das, S.; Attia, P. M.; Chueh, W. C.; Bazant, M. Z. Electrochemical Kinetics of SEI Growth on Carbon Black: Part II. Modeling. J. Electrochem. Soc. 2019, 166 (4), E107–E118. https://doi.org/10.1149/2.0241904jes.

[5] Attia, P. M.; Das, S.; Harris, S. J.; Bazant, M. Z.; Chueh, W. C. Electrochemical Kinetics of SEI Growth on Carbon Black: Part I. Experiments. J. Electrochem. Soc. 2019, 166 (4), E97–E106. https://doi.org/10.1149/2.0231904jes.

[6] Smith, R. B.; Bazant, M. Z. Multiphase Porous Electrode Theory, J. Electrochem. Soc. 2017, 164 (11). https://doi.org/10.1149/2.0171711jes.

[7] Bazant, M. Z., Theory of chemical kinetics and charge transfer based on nonequilibrium thermodynamics, *Accounts of Chemical Research*, *46*(5), 1144–1160. https://doi.org/10.1021/ar300145c

[8] T. Gao, Y. Han, S. Das, T. Zhou, D. Fraggedakis, N. Nadkarni, C. N. Yeh, W. Chueh, J. Li, M.Z. Bazant, Interplay of lithium intercalation and plating on graphite using in-situ optical measurements, submitted.

[9] Huang, W.; Attia, P. M.; Wang, H.; Renfrew, S. E.; Jin, N.; Das, S.; Zhang, Z.; Boyle, D. T.; Li, Y.; Bazant, M. Z.; McCloskey, B. D.; Chueh, W. C. and Cui, Y.; Nano Letters 2019 19 (8), 5140-5148. DOI: 10.1021/acs.nanolett.9b01515

589787 Model Development of Solid Oxide Fuel Cell Thermal Performance Using Artificial Neural Network

Abstract:

NASA's Johnson Space Center has recently begun efforts to eventually integrate air-independent Solid Oxide Fuel Cell systems, with landers that can be propelled by LOX-CH4, for long duration missions. Using landers that utilize such propellants, provides the opportunity to use such systems as a power option, especially since they will be able to process methane into a reactant through fuel reformation. To ensure fuel reformation in the systems, Steam Methane Reformation (SMR) will be employed. Various lead-up activities, such as hardware testing and computational modelling, have been initiated to assist with this developmental effort including the development of dynamic empirical models using artificial neural network. Such a modeling approach has shown to be accurate (R^2>0.9) to predict the thermal and electrical performance for SOFC systems. Hence, such models could be used for system optimization and control design.

Impact of Spacer Nature on Rheological Properties of Cationic Gemini Surfactant

Abstract ID: 589650

Shams Kalam[1], Muhammad Shahzad Kamal[2], Shirish Patil[1] and S. M. Shakil Hussain[2]

(1) KFUPM, Dhahran, Saudi Arabia

(2) Center for Integrative Petroleum Research, King Fahd University of Petroleum & Minerals, Dhahran, Saudi Arabia

Surfactants play a vital role in Chemical Enhanced Oil Recovery (cEOR) techniques. Gemini surfactants are a unique style of surfactants having more water-soluble hydrophilic head and oil-soluble hydrophobic tail units. Both groups are connected by a spacer at or near to the head groups. Polymers are added into the surfactants to increase the sweep efficiency of the flooding processes. This paper presents the rheological properties of Surfactant Polymer (SP) solutions in which two recently synthesized cationic gemini surfactants with different spacer nature are used.

The rheological properties of novel SP solutions were determined using Discovery hybrid rheometer (DHR-3) from TA Instrument. The geometric configuration set for this work was concentric cylinder. All experiments were carried out at 80 °C, while the shear rate for steady shear viscosity measurements ranges from 0.001 to 1000 s^{-1}. 0.25 weight % of a polymer FO 4290 SSH was used with different concentrations of newly synthesized gemini surfactants for Rheological measurements. Two hours magnetic stirring and twenty hours retention time were provided to make uniform surfactant polymer solutions.

Both surfactants exhibited good thermal stability in seawater with no precipitation. Shear viscosity and storage modulus were measured as a function of shear rate and angular frequency respectively at different surfactant concentrations ranging from 0 to 0.5 mM, where 0 mM shows pure polymer. We discovered that counter ions also influence the rheological properties significantly. Research outcomes revealed that the novel SP solution with a benzene ring and chlorine counterion produces better performance in comparison with benzene ring and bromine counterion. Moreover, the effect enhances if we increase one more benzene ring in the spacer.

This study reports two recently synthesized SP solutions for cEOR processes in carbonate reservoirs. It presents the effect of counterion on rheological properties and improves the understanding of tuning the surfactant structure in making SP solutions to increase the flooding sweep efficiency.

590257 Real Time Energy Management Systems Implementation for Continually Optimized Performance

Abstract:

It is well known that energy management is key for the optimal operation of refining and petrochemical sites. Industrial energy systems are constituted by complex utilities networks that are interconnected and have complex interactions that in order to achieve it best performance need to comply with increasing environmental rules, push against operating constraints and adapt to altering power prices and rapidly changing equipment availability.

After the initial installation, management and continuous improvement of these systems present challenges not only from a technological but also from an organizational point of view.

The implementation of an advanced digital Energy Management System (EMS) which includes capabilities for multi-period optimized scheduling, real time optimization and monitoring, deployed either as cloud-based or on-premise solutions that could be acting on open or closed loop is known to provides large benefits.

Critical is the sustainability of the EMS in the ever-evolving energy supply and demand of refining and petrochemical sites. Commercial, asset and organizational factors all change and the EMS needs to adapt to the developing environment of each enterprise meaning the type of implementation, including appropriate management structure, culture, best practices and Key Performance Indicators (KPIs) needs to be continually improved.

The EMS implementation at the large and complex Eastman site in Longview, TX, will be presented as an example. It will provide insight on the initial successes achieved in real time energy management for both, cost optimization and energy related KPIs performance tracking. The case study evolves to discuss how technical asset improvements, once replicated in the EMS, provided alternative operating modes, large additional benefits and sustained optimal

Andrew Lammers, NN2

Eastman Longview

Pablo Montagna*, Cristian Rojas

KBC Advanced Technologies, Inc

589818 Analysis of Fuel Cell System through Project Based Learning for Application of Classical Thermal Fluid Concepts in Mechanical Engineering Laboratory Course

Abstract:

A hands-on project involving analysis of air noncircular channels in Fuel Cell to measure the pressure was introduced in a thermal fluid laboratory course. Although the students studied fluid mechanics as a theoretical course, this project gave them the opportunity to build an experimental setup from scratch to help showcase their hands-on skills and to apply their understanding of relevant concepts of fluid sciences. The results of the project included a successful test of a conceptual procedure, designed to measure and calculate pressure drop and flow rates across the anode and cathode sides of a fuel cell. The students enjoyed completing and learning from the project. They also provided very useful feedback on the project by incorporating a functional experimental system for the future laboratory project. This project based learning opportunity allowed students to build their confidence in what they are learning as well as provide a glimpse of what the real world may hold.

589438 Measured Versus Calculated Sour Gas Properties

Abstract:

Measured versus calculated sour gas properties

Abstract:

Recent Environmental agencies have restricted standards for sulfur emission from gas chemical processing. Sour gas can extremely impact the delivery and processing by corroding the equipment. Gas properties such as compressibility, density and viscosity are a major factor controlling prediction of future production pipelines, and production tubing and pipelines in gas industry.

Sour gas effect on all gas industry aspects usually gas properties are measured in laboratories but occasionally experimental values are not available due to complexity and time consuming hence a reliable method would be proposed.

We are assessing the accuracy of the industry standard methods in calculating sour gas properties, compressibility factor using the following correlations:

Drunchuck - Abukkasem, Hall – Yarbarough, Drunchuck- Purvis- Robinson, Londono et-al, Alanazi et-al and Azizi et-al.

Viscosity using empirical correlations such as Carr-Kobayashi Burrows, Lea-Gonzalez –Eakin and Dean- Stiel.

Density using equations of state Redlich Kwang and Soave -Redlick - Kwang.

This study shows that the best correlation for calculating compressibility factor and density of high concentration sour gas is Londono-et al, all other considered methods are not accurate. All other methods investigated include high error especially for high H2S concentration.

However the error in estimation of Z factor for most of the considered methods has great impact on estimated sour gas density. Therefore we recommend that sour gas properties be measured experimentally at high pressure and high temperature conditions.

587967 The Study of a S-CO2 Oxy-Fuel Combustion Power Plant for Low Grade Coals

Abstract:

Recently, it is very important to have reliable solutions that can maximize power generation efficiency of current power generation processes and decrease emissions of CO2 and other pollutants because there are increasing concerns about the global warming and the depletion of fossil fuels. Among the available carbon capture technologies and power generation technologies, a S-CO2 (supercritical carbon dioxide) cycle based oxy-fuel combustion process appears to be a promising mid-term solution for the production of electricity from fossil fuels. Since supercritical carbon dioxide (S-CO2) involves a rapid density change at small temperature and pressure ranges, the S-CO2 cycle can extract large amounts of energy from relatively small equipment and it is possible to minimize or simplify a power generation block and to obtain higher thermal efficiency than existing steam cycles used in power generation plants. Therefore, an indirect S-CO2 oxy-fuel combustion power plants for low grade coals was designed and analyzed by using ASPEN-PLUS. Also, the operating characteristics of the indirect S-CO2 Oxy-fuel combustion power plant were compared with those of conventional air combustion power plants.

590099 Characterization and Performance Assessment of Oil Entrapment By Commercially Available Surfactants

Abstract:

Oil spills occur frequently and are a threat to the health and safety of human society and the well-being of the environment and ecosystem. Remediating oil spills is an ongoing challenge. Surfactant based dispersing agents, such as the Corexit® product line, are widely used for remediation of oil spills mainly through direct application on top of the oil slick. This work encompasses the investigation of variables that influence the emulsification process and therefore the encapsulation of oil within the surfactant micelles. Of central importance are the conditions under which oil entrapment in emulsions even occurs. A central finding is that there appears to be a threshold shear rate for emulsifying oil with surfactants, below which no emulsification occurs. Different schemes for applying shear stress in this work included simply stirring, using a high energy colloid mill, and flowing the mixture through a high shear capillary. Also, the sequence in which the oil/surfactant/water (O/S/W) components are mixed impinges strongly on whether and how much oil entrapment occurs. To try to correlate shear stress and velocity profiles throughout the sample volume with entrapment efficiency and kinetic stability of these O/S/W systems, Computational fluid dynamics (CFD) simulations were conducted. Another important factor is the ionic strength of the aqueous medium, which greatly impacts kinetics of these emulsions, but with a smaller impact on the threshold at which any emulsion is obtained at all. Surfactants tested include Corexit® EC9500A, sodium dodecyl sulfate (SDS), Triton X-100, and cetyltrimethyl ammonium bromide (CTAB). The quantification of oil entrapment was performed through either thermal gravimetric analysis (TGA) or differential scanning calorimetry (DSC).

Designing a Native Model for Assessment of the Effectiveness of Advertising

Sorush Niknamian
PhD, military Medicine Department, Liberty University, USA
Email: so.niknamian@gmail.com

Abstract

Reviewing and evaluating the effectiveness of advertising can be used as a criterion to convert weaknesses into strengths and budgeting. Also, Identification of the factors influencing the effectiveness of advertising can lead to more professional design and implementation. In the banking industry, with the increasing number of banks and increased right of customers to choose, the role of advertising has become increasingly important to succeed in this field of competition; so it should be said that advertising can make the bank close to its goals when it is sufficiently effective. The current research aimed to identify and rank the effective factors on the design of a native model for evaluation of the effectiveness of advertising of the Eghtesad Novin bank in Iran. The statistical population of this research is all managers and experts of the Eghtesad Novin bank. According to the research literature, the criteria and sub-criteria affecting the effectiveness of advertising have been extracted. By using the fuzzy Delphi technique and based on the experts' opinion, the final indices were extracted. Then, using the fuzzy AHP method, the rank of criteria has been addressed. The derived criteria are demographic characteristics, individual emotions, and attitudes of the customer, apparent features of the message, message content, market characteristics, cultural, economic features, design and nature of the service, advertising methods. The results of the ranking showed that the methods of advertisement, the design, and nature of the service, the apparent characteristics of the message, the emotions and attitudes of the customer, the demographic characteristics, the content of the message, economic, market characteristics, were placed in the first to ninth ranks, respectively.

Keywords: banking services, banking services marketing, advertising, advertising effectiveness,

Introduction:

Advertising is one of the promotional mix elements of marketing mixes, which play an important role in attracting customers. Advertising is a bridge between the organization and customers and can introduce the organization and its products to customers, inform consumers about the goods and services of the organization, convince them about the benefits of the product, and ensure them to have the best choice even after the purchase when the customer is uncertain (Buck, 2016: 172).

Given the increasing competition among the brands in the marketing sector, advertising is a necessary tool to introduce the characteristics of goods and services to customers. Leong et al. (1999) pointed out the fact that consumers overcome advertising information, and many of them do not remember the advertisements that they are exposed to, and it is necessary to consider the advertising and its attractive methods to attract customers' attention. The ability to attract customers' attention is one of the main indicators of advertising effectiveness that should be taken into consideration. The attention of customers is a key component of the overall effectiveness of advertising (Pileliene et al., 2017: 2). Any professional marketing without advertising will not be effective. Despite relatively high advertising costs, advertising is mentioned as an investment rather than the cost. Since many advertisements cause that the advertising companies do not benefit their advertising because of the wrong way of advertising, the competitive companies benefit from the advertising of that company (Hartnet, 2016: 23)

Advertising should be in line with the goals and plans, as well as the strategies of the organization. Therefore, we should see to what extent the advertising has met such goals. Allocating funds without evaluating the effectiveness of past advertising is the waste of resources of the organization. By evaluating the effectiveness of advertising, changes can be made to the form, the content of messages, type of media, budget, communication channels, and even the time and conditions for conducting advertising to create more useful and effective advertising than the past (Yazgirudo-Yasteh, 2015: 359). In the current competitive world, the importance of advertising is increasing. As a result, businesses have devoted a great deal of money to customers' attraction, but it is important to see whether the company advertising has been designed and implemented in such a way to meet the goals of the company. This issue has been often neglected in our country. (Kim, 2014: 2107).

The amazing effect of advertising on the lives of individuals requires that this scientific, social and communicative branch be carefully studied and advertising is made with a scientific, artistic, specialized and expert view. Professional advertisement can be a sharp sword for sales promotion, both inside and outside the country, and otherwise, it is a waste of time, money and energy. A professional advertising lever can, with the help of new technology, penetrate humans, and cause more desire for consumption (Arribeh, 2016: 463). Today, banks have turned their attention to the widespread use of mass media advertising, given the highly competitive structure governing on them. Also, given the relationship among a wide range of businesses of organizations and households and banks through various deposit accounts, loans, and other financial services, banks can benefit from advertising in a proper way to meet their goals, namely the introduction of their services, the communication with customers, their effectiveness on them in choosing a bank to do their banking business and ultimately attracting them. Unfortunately, in most banks, despite the massive advertising and the allocation of large amounts of energy and budget, the effectiveness of such advertisements is less evaluated. Since advertising is one-way, it is necessary to receive feedback from the customers (Rabiei, 2011: 19).

Regarding marketing, the effectiveness of advertising depends on the wide range of coverage of various consumers and the cost of advertising campaigns. Economists have defined the effectiveness of advertising as a change in the profit derived from promotional activities. Therefore, the effectiveness of bank advertising is the potential quality of information reported to consumers, and in promotional campaigns, it aims at creating a bank image and informing banking products and services, as well as obtaining a specific economic result before and after the transmission of the message (Denis et al., 2012: 85).

Eghtesad Novin Bank (EN bank), which is one of the oldest banks in Iran, due to the addition of new banks, should use new tools to protect and even increase its market share for the assessment of the effectiveness of its advertising. For this purpose and given that the bank uses various advertisements, it is necessary to design a model to evaluate the effectiveness of the bank advertising; therefore, the present study seeks to answer the following questions:

- What indicators are used to assess the bank native advertising model?

- How are the indicators for bank advertising evaluation weighed and ranked?

In this research, we have tried to identify the criteria using the Fuzzy Delphi process; then we will use the Fuzzy AHP method to rank the identified criteria. In the next section of the research, the literature has been studied. Then the research design has been expressed, and in the last section the conclusion has been presented.

1-Research literature:

1-1-Marketing of banking services

The growing trend in service activities has led to the emergence of new markets and large service companies. The complexity of service industries, such as information systems, management consultancy, and logistics, has increased, as service providers have entered the market with a variety of business backgrounds and capabilities to provide a broader range of services. Facing this complexity requires knowledge and skills to understand the services. The definition of the service includes all activities that have been done (for example, writing document specifications) (Selviaridis et al., 2013: 1398).

In the industrial literature, there are various ways to define the service and product requirements. Van Weele (2010) has distinguished technical and applied definitions from each other. The first requirement expresses the exact technical specifications of the goods with the required services and inputs while it focuses on performance demands and demand-oriented performance orientations. In this regard, Axelsson and Wynstra (2002) have proposed four methods for definition of services: input definitions are based on the sources and capabilities of the service provider while the process-oriented approach focuses on how services are presented regarding activities and processes. Output-oriented definitions focus on service capabilities and performance levels, as well as the interaction between the suppliers and recipients. Ultimately, they focus on stress-oriented results in the economic value of services (Selviaridis et al., 2013: 1399).

The financial sector forms an essential part of the service sector. In this regard, banks have used marketing to improve the strategic performance of banks. In particular, since the 1980s, this trend has been confirmed and criticized. Marketing researchers and experts are interested in researching the various aspects of marketing in banking. The quality of services is essential for customers who use banking services, and the provision of accurate and high-quality services leads to the survival of banks in the competitive market. Improving the quality of services will increase the access and response to the demands of the bank's customers, which will make the bank more distinct than other competitors. Therefore, measuring the quality of services in the banking industry is essential. Providing quality services is a hard and difficult task, in this regard, the features of services in general and banking services, in particular, should be explained. Banking services require a high degree of appropriate communication between service providers and customers. If services are not provided properly, they will be at high risk.

About the five features that distinguish them from material services (intangibility, non-allocation of materials for production, diversification, non-corruption, and lack of ownership), financial services have two other specific features, that have been firstly developed by McKechnie (1997). These characteristics include trusteeship and two-way information. These special features mean that the evaluation of services is more difficult than material products, and they not only create a challenge in the development of marketing activities, but also they are the source of the problem in providing an optimal level of quality services. Banking services require a high level of communication between service providers and customers, as well as a high degree of customer participation in the process of providing services. Angurr et al. (1999) emphasized that due to the high standardization and the development of technologies in the banking sector, human factor has been considered as the most important one in making distinctions among competitors. The expertise and motivation of the employees, as well as the level of customer involvement in the provision of services, significantly affect the quality of services. Due to the wide range of banking activities, attention should be paid to cultural and social changes that affect the attitudes of customers (Durdana et al., 2015: 100).

Marketing of banking services has three components as follows: if any of them were properly executed, banking services would change the country.

1- Market study: Identification of the banking services market system, including the status of current banking services and their weaknesses, strengths, expectations, demands, ideas and opinions of customers, the conditions of environmental instruments affecting the market system, in particular government policies and governing laws, and even global changes in the banking system and successful models in this system. Indeed, market study is the foundation of the evolution of the banking system, because market study (which is the result of constant communication with the market and conducting marketing studies and research in various forms) leads to comprehensive, up-to-date, relevant and necessary information for decision-makers and bank managers (Munsung Rheea & Satish Mehrab, 2006: 508).

2. Market creation: It involves the use of a variety of simple and complex marketing techniques and competitive strategies, as well as the growth, development and market creation in the banking system; it needs to understand the internal customers (employees) and external

customers, to communicate with them and meet their needs and demands. Market creation in the competitive environment requires appropriate strategies (Moch, 2013: 2908).

3. Market protection: One of the most important tasks of marketing in the banking system of the country is to maintain market share, a stable relationship with customers, and to convert ordinary customers to loyal and supportive ones. The more intense the competition, the more valuable the loyal customers will be, as marketing research in the world has shown that the cost of finding a new customer in firms is 3 to 33 times higher than the cost of maintaining current customers regarding the type of products and services. Therefore, market protection prevents the costly advertising, and attraction of new customers and provides a kind of confidence and serenity for market survival, especially in competitive conditions (Fredriksson & Moro, 2014: 69).

1-2-Effectiveness of bank advertising

One of the indicators of success in any organization is the extent to which the organization achieves its goals. Here, the success of any organization is determined by its own goals. This approach has been called effectiveness in the management literature. The effectiveness is the degree to which the organization achieves its goals. Goals must first be defined to measure effectiveness in any organization. Then, plans are set up to fulfill these goals, and thus the plans will be implemented. In the end, the results are compared with predetermined goals (Manjegani et al., 2015: 124).

About 100 years ago, William Heketh statement, "Half of the advertisements were wasted, we just do not know which half it is" attracted everyone to the accountability of advertising. What matters to advertisers is that how much the amount of money spent on advertising has returned? Has the cost spent on making and broadcasting the advertising been effective? Has the effect of advertising been strong or weak? Knowing the answer to these questions is important to the advertiser because today companies pay much money to make and broadcast advertising. According to Jack Trout, the money we use to advertise is very much. Every second of the World War II was $ 9,000, and Vietnam cost $ 22,000 per minute. But in the American Football Final, every minute of advertising cost $ 2 million, so the advertiser has the right to know about the effectiveness of his advertising (Aksoy, Atilla, 2005, Yeni Reklamcilik, Istanbul Bilgi Universitesi, 95). It is very difficult to measure the impact of long-term advertising on many occasions. Also, academically in the world of advertising, it is impossible to create a comprehensive theory that can be used to determine the effectiveness of advertisements on people at different levels under different circumstances. The effectiveness of media is its capacity to create desired results, or special conditions in a particular position and content, that is, attaining the desired communication goals (Wery, 2012)

Evaluating the effectiveness of advertising indicates how much advertising has succeeded in achieving the goals. The evaluation of the effectiveness of advertising acts as a feedback, informs us about the weaknesses and strengths of our advertising by which we find our shortcomings in the advertising. The marketer may choose an appropriate stage regarding product life or brand position, and use it to define his advertising goals and ultimately evaluate its effectiveness (Eckler & Bolis, 2011: 2).

The effectiveness of advertising is a function of its content (message), the execution (how it is transmitted), and repetition (the number of times each consumer sees the message) (Cutler, 1997, quoted by Moorthy & Hawkins, 2005). Some of the effectiveness criteria are the call and recognition of the brand name (Park, 2006), awareness, willingness and intention to purchase (Decrop, 2007), attitude towards the ad, attitude toward the brand name and intent to purchase (Muda et al., 2014)

The control of the communication element has a direct effect on customer perception of different dimensions of the brand name, such as service quality and loyalty. It means that the elements of communication, advertising and promotion must be carefully implemented and controlled (Suvittawat, 2015). Informative advertising is one of the factors that shape the behavior of the bank customers. Studies conducted both on attracting, retaining and loyalty of customers, show the need to examine the impact of advertising on attracting and loyalty of customers (Ioannou, Boukas and Skoufari, 2014: 25).

An ideal model of world-class advertising can: 1. transfer a consistent meaning to the consumer all over the world. 2. Transfer positive global capacity for the brand. That is, the results of advertising among different peoples should produce a positive effect and transfer high-quality signals (Strebinger et al., 2018: 410).

Advertisers want to prepare consumers to like goods or services, but sometimes it harms them through unwanted and intrusive advertising. It leads to the creation of a series of negative emotions in advertising. The purpose of ads is to create positive feelings in consumers about advertised goods and services (Bell & Buchner, 2018: 1).

On the other hand, research shows that commercial advertising can be positively evaluated and considered, and this positive evaluation leads to a positive evaluation of the brand with the product. However, advertising can have negative effects that lead to the formation of negative attitudes towards that product and service. Therefore, by increasing the volume of messages, it can have positive, negative or neutral effects. Therefore, marketers should evaluate advertising messages and their behavioral outcomes. With this regard, the effectiveness of advertising, based on a hierarchical approach, should be fully investigated (Tafani et al., 2018: 125).

There are various factors which influence effectiveness of advertisement like media, time, market life stage, advertisement life stage (new or old advertisement) and also advertisement creative signs (like content and administrative elements) (Talis et.al, 2005). Also using popular cartoon personages (Ranjbarian, 2009), size (Rousat, Venous & Ebrahimi, 2004, Baltas, 2003), color (Park, 2008, Moriarty, Mitchell, Wells, 2009, Cheng & Kao, 2011), complexity or simplicity of the advertisement (Morrison & Dainoff, 1972, Di Res, 2008), using irony factor (Mukherjee, Hober, 2004), using famous artists (Rousta, Venous, Ebrahimi , 2004) and consumer features (Matrin, Phaymi, Agee, 2002, Park, 2008) are among these variables. Advertisement repetition is also a factor which helps in reminding its content to the audiences (Rousta, Venous, Ebrahimi, 2004). One of considered factors in the debate of evaluating effectiveness of advertisement is the attitude towards advertisement. Attitude towards the advertisement includes public interest or lack of interest in an advertising stimulus in encountering a special topic (Moaen & Minor, 2009). Mental imagery (B one & Elen, 1992), content transparence and emotions called through the advertisement are among variables influencing the attitude towards the advertisement. Some of these factors may lead to strong impacts while some of them may create weak impacts or don't include any impact. Awareness of the impact of each of these factors makes a considerable help in making management decisions (Talis et.al, 2005).

Fotini et.al (2009) provided a model of effective advertisements for advertising websites of big companies. The results of the study showed that measurements, activities, priorities and intentions of people have a considerable impact on evaluation and qualitative factors of relation-oriented marketing have a great impact on advertisements effectiveness indicators. The extracted criteria include:

- Advertisement policies: product information, company information, product general display, consistency with traditional advertisement

- Medium features: message features, information structure, responding time, facilitating research, beauty, interactions, communications in real-time, amusement, information control

- Relation- oriented marketing policy: communications, feedback, supporting customers

Today, organizations for competition in the business world and selling their goods and services require purposeful and efficient marketing activities. One of topics which has a very important role in the process of marketing of an organization and assign a high volume of budget to it is advertisement. Advertisement causes some changes in the knowledge, attitude and behavior level of audiences, so attention to evaluating advertisement effectiveness is a necessity for the organizations that consider advertisement as requirement of continuous life, growth and development. By assessing advertisement effectiveness and investigating consistency with the organization objectives and its expected impacts, some changes may be created in its dimensions, content and methods so that a more effective and useful advertising is performed. Regarding the research literature and internal and external background which has been stated, here a general categorization of the model proposed criteria has been provided in the following table:

Table 1- summation and extracted variables

scholar	sub standard	criterion
) Dargi & Mohammadian (2010), Ashouri (2016), Mansouri Moayed (2005), Aghapour (2016), Mirzadh & Mehrmanesh (2016), Keramforn (2011), Mac Ki- Nesbiat et.al (2011), Rosemary et.al (2012), Wang (2015), Maryion et.al(2017)	awareness of sexuality differences	individual factors
Dargi & Mohammadian (2010), Ashouri (2016), Samsam Shariat Keramforn (2011)(2007), Mirzadeh & Mehrmanesh (2016), Mac Ki- Nesbiat et.al (2011), Rosemary et.al (2012), Wang (2015), Maryion et.al(2017)	attention to age requirements	
Dargi & Mohammadian (2010), Dehdashti Shahrokh (2009)	various consumers (permanent, ordinary and case consumers)	

Khashizadeh (2010), Danaei & Dasar (2016), Pillay et.al (2012), Marion et.al (2017)	advertisement richness	
Danaei & Dasar (2016), Dargi & Mohammadian (2010), Pillay et.al (2012), Olson et.al (2012)	advertisement accuracy	
Ashouri (2016), Rabiei (2011), Taeini Harari (2014)	positive mentality of the organization	
Dargi & Mohammadian (2010), Shahmohammadi (2011), Olson et.al (2012)	honesty feeling in advertisement	
Fakhari et.al (2012)	threatening feeling	
Ashourizadeh (2016), Hamidizadeh (2013)	attitude towards brand	
madhoushi (2013), Olson et.al (2012)	being convinced	
Ahmadi (2013), Mac Ki – Nesbiat et.al (2011), Olson et.al (2012)	convicting audiences	
Ashouri (2016), Hosivar (2014), Olson et.al (2012)	mental imagery	
Siadat & Sakhi Jourashk (2014)	positive feeling	
Navaei et.al (2015), Rozmari et.al (2012)	audience being active	
Mosavi Esmaeilzadeh (2015), Olson et.al (2012)	knowledge level	
Dargi & Mohammadian (2010), Ashouri (2016), Shafiei Nikabadi & Zarei (2017)	attitude towards advertisement	
Dargi & Mohammadian (2010), Hosseinzadeh et.al (2016), Sanei et.al (2016)	familiarity with methods and forms of advertisement	message
Ashouri (2016), Olson et.al (2012), Alan (2012), Patric Ro (2014)	attention to reduction of expectation time of the user for observing advertisement	
Danaei & Dasar (2016), Forentebury & Mac Gldric (2011)	beautiful and appropriate	

	combination of colors	
Aghapuor (2016), Sanei et.al (2016)	site and location of advertisement placement	
Hassanzadeh & Danaei (2016), Sanei et.al (2016)	access to sufficient and required information	
Dargi & Mohammadian (2010), Darzban (2016)	starting the advertisement with a challenge	
Navaei et.al (2015), Salimi Fard & Pourdehghan (2016), Bargar et.al (2016)	starting the advertisement with a statement and explanation	
Dargi & Mohammadian (2010), Mirzadeh & Mehrmanesh	using helping words and terms for decision making	
Dargi & Mohammadian (2010), Daneshvar et.al (2016), Alan (2012)	using stimuli with diversions more than usual and expected extent	
Sanei et.al (2016)	using images of women and children	
Sohrabi Ranani (2016), Fortenbury & Mac Goldric (2011)	creativity in the message	
Shafei Nikabadi & Zarei (2017), Denis et.al (2012)	the text of advertisement message being a predicate sentence	

Khashizadeh (2010), Dargi & Mohammadian (2010), Sohrabi Ranani (2016), Patric Ro (2014), Abdolsalam et.al (2016)	continuation and repetition of message	
Daneshfar & Nabizadeh (2015), Fortenbury & Mac Goldric (2011)	using trustable and famous people	
Salimifard & Pourdehghan (2016)	common dialects and terms	
Daneshfar & Nabizadeh(2015), Forentnburi & Mac Goldric (2011)	loveliness of the popular person for the observer	
Daneshfar & Nabizadeh (2015), Forentnburi & Mac Goldric)(2011	familiarity of famous person with the observer	
Mansouri Moayed (2005), Aghapour (2016)	emotional music	
Khonsazadeh (2010), Hosseinzadeh et.al (2016), Aghapour (2016)	using various methods	
Siadat & Sakhi Jourashk (2014), Sohrabi Ranani (2016)	high coverage and being universal	
Navaei et.al (2016), Mosavi Esmaeilzadeh (2015), Hosseinzadeh et.al (2016), Mirzadeh & Mehrmanesh (2016), Rosemary et.al (2012), Mikit et.al (2012), Spenkar (2013)	media advertisement	
Navaei et.al (2015), Mosavi Esmaeilzadeh (2015), Hosseinzadeh et.al (2016), Ashouri (2016), Hassanzadeh & Danaei (2016), Mikit et.al (2012)	multimedia advertisement	
Navaei et.al (2015), Mirzadeh & Mehrmanesh (2016)	radio advertisement	
Navaei et.al (2015), Mosavi Esmaeilzdeh (2015), Ashouri (2016), Hassanzadeh & Danaei (2016)	cell phone advertisement	
Navaei et.al (2015), Hosseinzadeh et.al (2016), Mirzadeh & Mehrmanesh (2016)	creating TV advertisement	
Ahmadi (2013), Salimi Fard & Pourdehghan (2016), Hoang et.al (2013)	attraction of rewards	
siadat & Sakhi Jourashk (2014), Koumra et.al (2016)	being benevolent	

Darzaian (2016), Zabkouich & Louksten (2011), Abdolsalam et.al (2016)	coordination in speaking and behavior	
Ahmadi (2013), Salimi Fard & Pourdehghan(2016), Honari et.al (2015)	reality of messages	
Madhoushi (2013), Salimi Fard & Pourdehghan (2016), Houang et.al (2013)	graphic-designing	
Shahmohammadi (2011), Madhoushi (2013), Salimi Fard & Pourdehghan(2016)	psychological – geographical division	
Mirzadeh & Mehrmanesh (2016), Zabkouich & Lokesten (2011), Menroz et.al (2012), Choubhai et.al (2013), Mokan & Oltz (2014), Ayono et.al (2014)	advertisement on web	
Dehdashti Shahrokh (2009), Shahmohammadi (2011), Mirzadeh & Mehrmanesh (2016), Menroz et.al (2012), Choubhai et.al (2013), Mokan & Oltz (2014), Ayono et.al (2014), Henri et.al (2015), Yazghiro- Basteh (2015), Lin (2016)	internet & cell phone advertisement	
Shahmohammadi (2011), Sanei et.al (2016), Menroz et.al (2012), Chobai et.al (2013), Ayono et.al (2014), Henri et.al (2015)	selecting advertising media	
Shahmohammadi (2011), Sanei et.al (2016), Howang et.al (2013), Henri et.al (2015), Yazghirdo-Yasteh (2015), Martina Santa et.al (2017)	artistic aspect of advertisement	
Kashizadeh (2010), Hoshivar(2014), Pelpin et.al (2017)	gender of speaker in the advertisement	
Hosseinzadeh et.al (2016), oribeh (2016), Madra et.al (2017)	text repetition	
Shahmohammadi (2011), Hassanzadeh & Danaei (2016), Abri & Joumar (2014), Ayouno et.al (2014), Oribeh (2016), Pachman (2016), Bak (2016), Madra et.al (2017)	advertisement	
Samsam Shariat (2007), Sohrabi Ranani (2016), Daneshvar et.al (2016), Bak (2016), Kiji-Kali (2017)	factors relating to usage	environment

Fakhari et.al (2012), Mirzadeh & Mehrmanesh (2016), Howang et.al (2013), Bak (2016), kiji-Kali (2017)	factors relating to facilitation	
Dargi & Mohammadian (2010), Hamidizadeh (2013), Darzban (2016), Belman et.al (2012)	awareness of different status of the environment status	
Dargi & Mohammadian (2010), Mirzadeh & Mehrmanesh (2016)	awareness of access location of objective group to the product	
Dargi & Mohammadian (2010), Salimifard & Pourdehghan (2016)	awareness of norms	
Dargi & Mohammadian (2010), Hamidizadeh (2013), Darzban (2016), Belman et.al (2012), Koumra et.al (2016)	awareness of religions and religious values	
Dargi & Mohammadian (2010), Hassanzadeh & Danaei (2016)	awareness of method of evaluating services in relation to other services	
Hassazadeh & Danaei (2016), Radmiker (2015)	attention to security	
Samsam Shariat (2007), Mirzadeh & Mehrmanesh (2016), Belman et.al (2012)	economic beliefs	
Salimifard & Pourdehghan (2016), Pilay et.al (2012), Howang et.al (2013), Komra et.al (2016)	ethical beliefs	
Salimifard & Pourdehghan (2016)	Islamic teaching	
Samsam Shariat (2007), Daneshvar et.al (2016), Radmiker (2015)	strategic look to advertisement plan	
Dargi & Mohamadian (2010), Ansari & Rasei (2016), Abri & Joumar (2014)	market share	
Dargi & Mohamadian (2010), Darzain (2016)	obviating primary needs of audiences	

Salimi Fard & Pourdehghan (2016), Kim et.al (2012), Howang et.al (2013)	reference groups	
Sohrabi Ranin (2016), Fontini et.al (2009), Belman et.al (2012)	awareness of services	factors relating to services
Salimi Fard & Pourdehghan (2016), Abri & Joumar (2014), Radmiker (2015)	stating potential benefits of purchase ad services consumption	
Dargi & Mohamadian (2010), Hosseinzadeh et.al (2016), Fontini et.al (2009)	rate of services complexity	
Samsam Shariat (2007), Darzaian (2016), Fontini et.al (2009), Radmiker (2015)	display of expected experience	
Khashizadeh (2010), Darzain (2016)	rate of familiarity of consumer with services	
Ansari & Rasaei (2016), Belman et.al (2012)	market researches	
Hosseinzadeh et.al (2016), Mirzadeh & Mehrmanesh (2016)	record of exploiting services	
Dargi & Mohammadian (2010), Ansari & Rasaei (2016), Lousi (2012), Park (2015), Marion et.al (2017)	being unique	
Dehdashti Shahrokh (2009), Daneshvar et.al (2016), Radmiker (2015), Jankoski et.al (2016)	price, benefits	
Daneshvar et.al (2016), Fontini et.al (2009), Jankoski et.al (2016)	relation with customer	
Samsam Shariat (2007), Daneshvar et.al (201), Fontini et.al (2009), Jankoski (2016)	supporting customer	
Dehdashti Shahrokh (2009), Ansari & Rasaei (2016), Fontini et.al (2009), Radmiker (2015), Jankoski (2016)	feedback	

2- Research plan

The present study is of descriptive applied type which has been performed with survey method. The statistical population of this study includes all managers and experts of Eghtesad

Novin Bank. The statistical population of the present study contains all people who are exposed to advertisements of Eghtesad Novin Bank across the country. For selecting samples, Cochran formula is used for unlimited societies that based on this formula, the samples number were determined 400 persons for each province. The sample people were selected based on the following sampling method. At first, using stratified sampling method , the samples were randomly selected from five geographical regions of the country (north, south, east, west and center) of nine provinces (Mazandaran, Khorasan Razavi, Zanjan, Chahar Mahal Bakhtiari , Kerman, East Azarbaijan , Kermanshah, Ilam and Tehran).

3- Variables and research model

In this study, fuzzy Delphi technique has been used for identifying main indicators. For ranking the final criteria and sub standards, multi-criteria decision making models and AHP model have been used with fuzzy approach. For analysis of obtained data, Excel software and coding in the environment of this software have been used. The results obtained from data analysis have been provided in the present chapter.

3-1 Identifying criteria and sub standards of factors effective on advertisement effectiveness

At the first step, the primary indicators were identified and selected. In the first stage, by reviewing the literature and performed studies, a large number of indicators were extracted from the literature. For screening the indicators and identifying the final indicators, fuzzy Delphi approach has been used. The expert views have been collected in the field of rate of significance of main indicators. Fuzzy average and defuzzification output of rates relating to indicators are presented in table 2. The rate of defuzzification greater than 7 is acceptable and every indicator which has the score higher than 7 is confirmed (Wu& Fang, 2011).

Table 2- fuzzy average and fuzzy screening of indicators of factors effective on advertisement effectiveness

Result	Crisp	mean	U	M	L	
acceptance	7.15	(2, 7.97, 9)	9	7.97	2	Q01
acceptance	7.84	(6, 8.01, 9)	9	8.01	6	Q02

Result	Crisp	mean	U	M	L	
acceptance	7.92	(6, 8.13, 9)	9	8.13	6	Q03
acceptance	7.39	(4, 7.83, 9)	9	7.83	4	Q04
acceptance	7.70	(6, 7.8, 9)	9	7.80	6	Q05
rejection	6.68	(2, 7.28, 9)	9	7.28	2	Q06
acceptance	7.71	(6, 7.82, 9)	9	7.82	6	Q07
acceptance	7.66	(6, 7.75, 9)	9	7.75	6	Q08
acceptance	7.02	(2, 7.78, 9)	9	7.78	2	Q09
acceptance	7.27	(2, 8.15, 9)	9	8.15	2	Q10
acceptance	7.09	(2, 7.88, 9)	9	7.88	2	Q11
rejection	6.85	(2, 7.52, 9)	9	7.52	2	Q12
acceptance	7.43	(4, 7.9, 9)	9	7.90	4	Q13
rejection	6.71	(3, 7.07, 9)	9	7.07	3	Q14
acceptance	7.17	(2, 8.01, 9)	9	8.01	2	Q15
acceptance	7.13	(2, 7.95, 9)	9	7.95	2	Q16
rejection	6.85	(4, 7.02, 9)	9	7.02	4	Q17
acceptance	7.43	(4, 7.89, 9)	9	7.89	4	Q18
acceptance	7.70	(6, 7.8, 9)	9	7.80	6	Q19
rejection	6.80	(3, 7.21, 9)	9	7.21	3	Q20
acceptance	7.75	(6, 7.88, 9)	9	7.88	6	Q21
rejection	6.62	(3, 6.93, 9)	9	6.93	3	Q22
acceptance	7.61	(6, 7.66, 9)	9	7.66	6	Q23
acceptance	7.22	(2, 8.08, 9)	9	8.08	2	Q24
acceptance	7.04	(2, 7.81, 9)	9	7.81	2	Q25
acceptance	7.71	(6, 7.82, 9)	9	7.82	6	Q26
rejection	6.65	(3, 6.97, 9)	9	6.97	3	Q27
acceptance	7.98	(6, 8.22, 9)	9	8.22	6	Q28
acceptance	7.70	(6, 7.79, 9)	9	7.79	6	Q29
rejection	6.54	(3, 6.81, 9)	9	6.81	3	Q30
acceptance	7.19	(4, 7.53, 9)	9	7.53	4	Q31
acceptance	7.79	(5, 8.18, 9)	9	8.18	5	Q32

Result	Crisp	mean	U	M	L	
acceptance	7.53	(4, 8.05, 9)	9	8.05	4	Q33
acceptance	7.05	(2, 7.82, 9)	9	7.82	2	Q34
acceptance	7.43	(4, 7.89, 9)	9	7.89	4	Q35
rejection	6.92	(3, 7.38, 9)	9	7.38	3	Q36
acceptance	7.02	(2, 7.78, 9)	9	7.78	2	Q37
rejection	6.91	(3, 7.37, 9)	9	7.37	3	Q38
rejection	6.96	(2, 7.68, 9)	9	7.68	2	Q39
rejection	6.99	(2, 7.74, 9)	9	7.74	2	Q40
acceptance	7.42	(4, 7.88, 9)	9	7.88	4	Q41
acceptance	7.10	(2, 7.9, 9)	9	7.90	2	Q42
rejection	6.59	(2, 7.13, 9)	9	7.13	2	Q43
rejection	6.55	(2, 7.08, 9)	9	7.08	2	Q44
acceptance	7.77	(6, 7.9, 9)	9	7.90	6	Q45
rejection	6.91	(2, 7.62, 9)	9	7.62	2	Q46
acceptance	7.09	(2, 7.88, 9)	9	7.88	2	Q47
acceptance	7.68	(6, 7.78, 9)	9	7.78	6	Q48
acceptance	8.19	(6, 8.53, 9)	9	8.53	6	Q49
rejection	6.88	(2, 7.57, 9)	9	7.57	2	Q50
acceptance	7.75	(5, 8.12, 9)	9	8.12	5	Q51
acceptance	7.03	(2, 7.8, 9)	9	7.80	2	Q52
rejection	6.96	(4, 7.19, 9)	9	7.19	4	Q53
acceptance	7.04	(2, 7.8, 9)	9	7.80	2	Q54
rejection	6.96	(3, 7.45, 9)	9	7.45	3	Q55
acceptance	7.02	(2, 7.78, 9)	9	7.78	2	Q56
acceptance	7.92	(5, 8.38, 9)	9	8.38	5	Q57
acceptance	7.00	(2, 7.76, 9)	9	7.76	2	Q58
rejection	6.90	(1, 7.86, 9)	9	7.86	1	Q59
acceptance	7.10	(4, 7.4, 9)	9	7.40	4	Q60
acceptance	7.86	(6, 8.04, 9)	9	8.04	6	Q61
rejection	6.97	(2, 7.71, 9)	9	7.71	2	Q62

Result	Crisp	mean	U	M	L	
acceptance	7.02	(2, 7.79, 9)	9	7.79	2	Q63
acceptance	7.73	(6, 7.84, 9)	9	7.84	6	Q64
acceptance	7.01	(2, 7.76, 9)	9	7.76	2	Q65
rejection	6.65	(3, 6.98, 9)	9	6.98	3	Q66
acceptance	7.01	(2, 7.77, 9)	9	7.77	2	Q67
acceptance	7.90	(6, 8.11, 9)	9	8.11	6	Q68
acceptance	7.55	(5, 7.83, 9)	9	7.83	5	Q69
acceptance	8.02	(6, 8.28, 9)	9	8.28	6	Q70
acceptance	7.95	(6, 8.17, 9)	9	8.17	6	Q71
acceptance	7.03	(2, 7.79, 9)	9	7.79	2	Q72
acceptance	7.65	(5, 7.98, 9)	9	7.98	5	Q73
acceptance	7.00	(1, 8, 9)	9	8.00	1	Q74
acceptance	7.04	(2, 7.8, 9)	9	7.80	2	Q75
acceptance	7.03	(2, 7.8, 9)	9	7.80	2	Q76
acceptance	7.30	(4, 7.7, 9)	9	7.70	4	Q77
rejection	6.88	(2, 7.57, 9)	9	7.57	2	Q78
acceptance	7.63	(4, 8.19, 9)	9	8.19	4	Q79
acceptance	7.65	(5, 7.98, 9)	9	7.98	5	Q80
rejection	6.59	(3, 6.88, 9)	9	6.88	3	Q81
acceptance	7.09	(2, 7.88, 9)	9	7.88	2	Q82
rejection	6.83	(2, 7.5, 9)	9	7.50	2	Q83
acceptance	7.12	(2, 7.93, 9)	9	7.93	2	Q84

From the 84 indicators which were assessed, 24 indicators which have acquired the average score less than 7 are omitted. So, 60 indicators are finally confirmed (Adelazaret.al, 1995).

The remained indicators will be finally examined in the format of a 60-question questionnaire with analytical hierarchy process.

3-2 determining the priority of factors effective on effectiveness of advertisement using fuzzy AHP technique

For determining the priority of major criteria, fuzzy analytical hierarchy (FAHP) has been used. The analysis process is as follows.

1- Dual comparison of major criteria based on objective and determining the weight of major criteria

2- Dual comparison of sub standards of each criterion and determining sub standards of each cluster

3- Multiplication of the weight of sub standards in the weight of respective criterion and determining the final weight of sub standards

For dual comparison of elements, 9-degree Saaty scale has been used. 9-degree Saaty scale has been provided by Thomas Saaty , codifier of analytical hierarchy theory. Also, in this study fuzzy approach has been used for making the values quantitative. So, Saaty fuzzy range has been used.

Table 3- fuzzy range equal to 9-degree Saaty scale in AHP technique. Habibi et.al , 2014, 77

reversed fuzzy equivalent	fuzzy equivale nt	the word term of comparison status of i to j
(1,1,1)	(1, 1, 1)	Preferred Equally
$\left(\frac{1}{3},\frac{1}{2},1\right)$	(1, 2, 3)	intermediate
$\left(\frac{1}{4},\frac{1}{3},\frac{1}{2}\right)$	(2, 3, 4)	Preferred moderately
$\left(\frac{1}{5},\frac{1}{4},\frac{1}{3}\right)$	(3, 4, 5)	intermediate
$\left(\frac{1}{6},\frac{1}{5},\frac{1}{4}\right)$	(4, 5, 6)	Preferred Strongly
$\left(\frac{1}{7},\frac{1}{6},\frac{1}{5}\right)$	(5, 6, 7)	intermediate
$\left(\frac{1}{8},\frac{1}{7},\frac{1}{6}\right)$	(6, 7, 8)	very strongly Preferred
$\left(\frac{1}{9},\frac{1}{8},\frac{1}{7}\right)$	(7, 8, 9)	intermediate
$\left(\frac{1}{9},\frac{1}{9},\frac{1}{9}\right)$	(9, 9, 9)	Extremely Preferred

(source in papersL Khorshid & Zabihi, Lee my H.I; Wen-Chin Chen, Ching-Jan, Chang, 2008, 101)

The pattern of criteria and sub standards has been drawn in the figure using AHP technique. Also, the criteria and sub standards of the study have been named with numeric index so that the research flow is easily detectable.

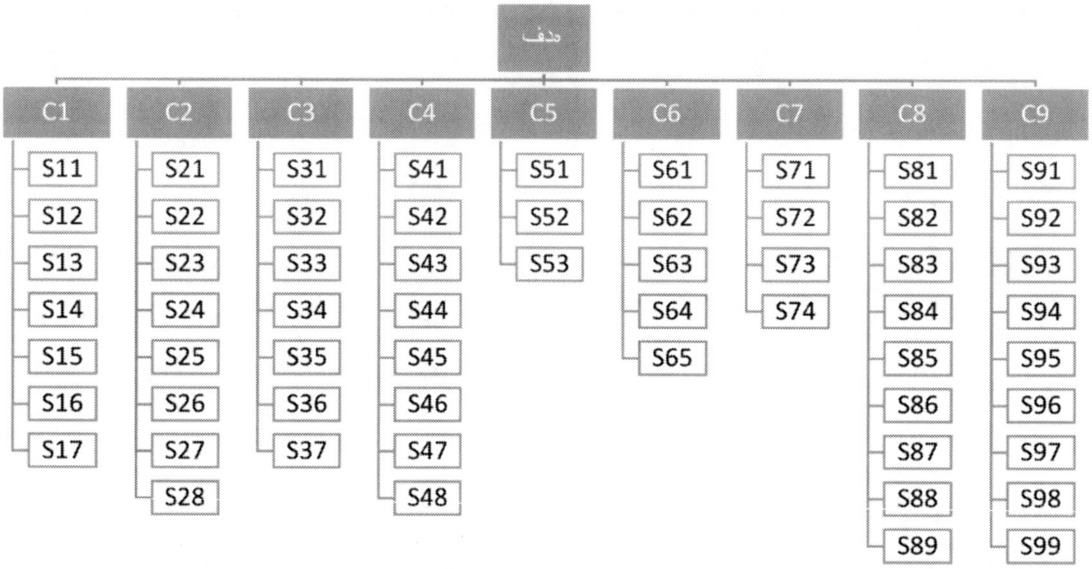

Figure 1- hierarchical model of criteria and sub standards of factor effective on advertisement effectiveness

1- Determining the priority of major criteria based on objective

In the first step, the major criteria based on the objective have been compared dually. Dual comparison is very simple and all elements of each cluster should be compared two by two. Therefore, if there is n element in a cluster, $\frac{n(n-1)}{2}$ comparisons will be performed. Since there are four criteria, so the number of performed comparisons is equal to:

$$\frac{n(n-1)}{2} = \frac{9(9-1)}{2} = 36$$

So, 36 dual comparisons have been performed from group perspective of experts. The experts view has been made quantitative using fuzzy scale. Collecting experts view has been collected with 9-degree Saaty range. So, 6 dual scales of four major four criteria have been performed based on the experts view. Using geometric average the experts views have been aggregated. Fuzzy aggregation will be calculated as follows:

Equation 1

$$F_{AGR} = (\prod(l), \prod(m), \prod(u))$$

After matrix formation of the obtained dual comparisons, the special vector has been calculated. At first, the fuzzy extension of each row is calculated. Every entry of dual comparison matrix $\tilde{X}$

Is displayed as $\tilde{x}_{ij}$.

The fuzzy extension of every row is also displayed with symbol $\tilde{S}_i$. Therefore, the fuzzy extension of each row is calculated as follows:

Equation 2

$$\tilde{S}_i = \sum_{j=1}^{n} x_{ij}$$

Then, the fuzzy extension of every row elements will be as follows:

Fuzzy extension of row 1 (1,1,1) ⊕ (1,27,1,47,1,75) ⊕ (0.47, 0.55, 0.66) ⊕ (1.52, 1.81, 2.14) ⊕ (0.93, 1.14, 1.39) ⊕ (1.26, 1.61, 2.04) ⊕(1.25, 1.51, 1.84) ⊕(0.78, 0.97, 1.24) ⊕ (0.59, 0.75, 0.96) ⊕ (9.08, 10.8, 13.02)

Fuzzy extension of row 2 (0.57, 0.68, 0.79)⊕(1, 1, 1)⊕(0.58, 0.7, 0.89)⊕(1.84, 2.28, 2.75)⊕(0.68, 0.82, 1.02)⊕(1.42, 1.81, 2.34)⊕(1.64, 2.04, 2.53)⊕(0.71, 0.88, 1.12)⊕(0.73, fuzzy extension of row 3 (1.52, 1.82, 2.11)⊕(1.13,)0.9, 1.1)⊕ = (9.17, 11.11, 13.55 1.42, 1.72)⊕(1, 1, 1)⊕(1.89, 2.34, 2.88)⊕(0.7, 0.86, 1.09)⊕(1.45, 1.78, 2.22)⊕(1.55, 1.88, 2.29)⊕(0.57, 0.71, 0.93)⊕(0.67, 0.8, 0.96)⊕ = (10.47, 12.61, 15.2)

Fuzzy extension of row 4 (0.47, 0.55, 0.66)⊕(0.36, 0.44, 0.54)⊕(0.35, 0.43, 0.53)⊕(1, 1, 1)⊕(0.97, 1.21, 1.51)⊕(0.96, 1.33, 1.8)⊕(2.14, 2.7, 3.37)⊕(0.94, 1.21, 1.56)⊕(1.3, 1.69, 2.16)⊕ = (8.49, 10.57, 13.14)

Fuzzy extension of row 5 (0.72, 0.88, 1.07) ⊕ (0.98, 1.22, 1.48)⊕(0.92, 1.17, 1.43)⊕(0.66, 0.82, 1.03)⊕(1, 1, 1)⊕(0.67, 0.83, 1.04)⊕(0.23, 0.29, 0.36)⊕(0.28, 0.32, 0.39)⊕(0.3, 0.38, 0.48)⊕ = (5.76, 6.91, 8.29)

Fuzzy extension of row 6 (0.49, 0.62, 0.79)⊕(0.43, 0.55, 0.7)⊕(0.45, 0.56, 0.69)⊕(0.55, 0.75, 1.04)⊕(0.96, 1.2, 1.48)⊕(1, 1, 1)⊕(0.34, 0.41, 0.51)⊕(0.27, 0.34, 0.44)⊕(0.24, 0.31, 0.41)⊕ = (4.74, 5.75, 7.08)

Fuzzy extension of row 7 (0.54, 0.66, 0.8)⊕(0.4, 0.49, 0.61)⊕(0.44, 0.53, 0.64)⊕(0.3, 0.37, 0.47)⊕(2.81, 3.5, 4.3)⊕(1.95, 2.43, 2.96)⊕(1, 1, 1)⊕(0.34, 0.42, 0.54)⊕(0.29, 0.36, 0.47)⊕ = (8.06, 9.76, 11.78)

Fuzzy extension of row 8 (0.8, 1.03, 1.28)⊕(0.89, 1.14, 1.41)⊕(1.08, 1.41, 1.76)⊕(0.64, 0.82, 1.06)⊕(2.55, 3.1, 3.63)⊕(2.25, 2.97, 3.72)⊕(1.86, 2.39, 2.94)⊕(1, 1, 1)⊕(0.58, 0.69, 0.83)⊕ = (11.66, 14.56, 17.64)

Fuzzy extension of row 9 (1.04, 1.33, 1.69)⊕(0.91, 1.11, 1.37)⊕(1.04, 1.25, 1.5)⊕(0.46, 0.59, 0.77)⊕(2.08, 2.63, 3.3)⊕(2.42, 3.23, 4.1)⊕(2.15, 2.74, 3.43)⊕(1.2, 1.46, 1.72)⊕(1, 1,)1)⊕ = (12.29, 15.35, 18.88

Then, the fuzzy sum of total priorities column elements is calculated

Equation 3

$$\sum \tilde{S}_i = \sum_{i=1}^{n} \sum_{j=1}^{n} x_{ij}$$

The sum of elements of the priorities column of major criteria will be as follows:

$$\sum_{i=1}^{4} \sum_{j=1}^{4} M_g^j = (79.72, 97.41, 118.59)$$

For normalization of each criterion priorities, that criterion total rates sum should be divided into sum of all priorities (column elements), since the rates are fuzzy, so fuzzy sum of every row is multiplied in summation reverse. The summation reverse should be calculated.

Equation 4

$$if\ \tilde{F} = (l, m, u)\ then\ \tilde{F}^{-1} = \left(\frac{1}{u}, \frac{1}{m}, \frac{1}{l}\right)$$

Then, based on the equation 4, we will have

$$\left(\sum_{i=1}^{n} \sum_{j=1}^{n} M_g^j\right)^{-1} = (0.01, 0.01, 0.01)$$

$$P_k = \sum_{i=1}^{n} M * \left(\sum_{i=1}^{n} \sum_{j=1}^{n} M_g^j\right)^{-1}$$

So, the results obtained from normalization of obtained rates will be as follows:

$\widetilde{W}_{C1}$ = (0.08, 0.11, 0.16)

$\widetilde{W}_{C2}$ = (0.08, 0.11, 0.17)

$\widetilde{W}_{C3}$ = (0.09, 0.13, 0.19)

$\widetilde{W}_{C4}$ = (0.07, 0.11, 0.16)

$\widetilde{W}_{C5}$ = (0.05, 0.07, 0.1)

$\widetilde{W}_{C6} =$ (0.04, 0.06, 0.09)

$\widetilde{W}_{C7} =$ (0.07, 0.1, 0.15)

$\widetilde{W}_{C8} =$ (0.1, 0.15, 0.22)

$\widetilde{W}_{C9} =$ (0.1, 0.16, 0.24)

Each of the obtained rates is fuzzy and normalized weight relating to major criteria. There are various methods for defuzzifcation of the obtained rates. In this study, Bojadziev proposed method has been used for defuzzification.

Table 4- defuzzification and computation of final and normal weight of criteria of factors effective on advertisement effectiveness

Normal	Deffuzy	X3max	X2max	X1max	Crisp
0/111	0/117	0/113	0/115	0/117	C1
0/114	0/120	0/115	0/118	0/120	C2
0/103	0/137	0/133	0/135	0/137	C3
0/108	0/113	0/112	0/113	0/113	C4
0/070	0/073	0/072	0/073	0/073	C5
0/060	0/063	0/062	0/063	0/063	C6
0.101	0/107	0/103	0/105	0/107	C7
0.149	0/157	0/153	0/155	0/157	C8
0.158	0/167	0/163	0/165	0/167	C9

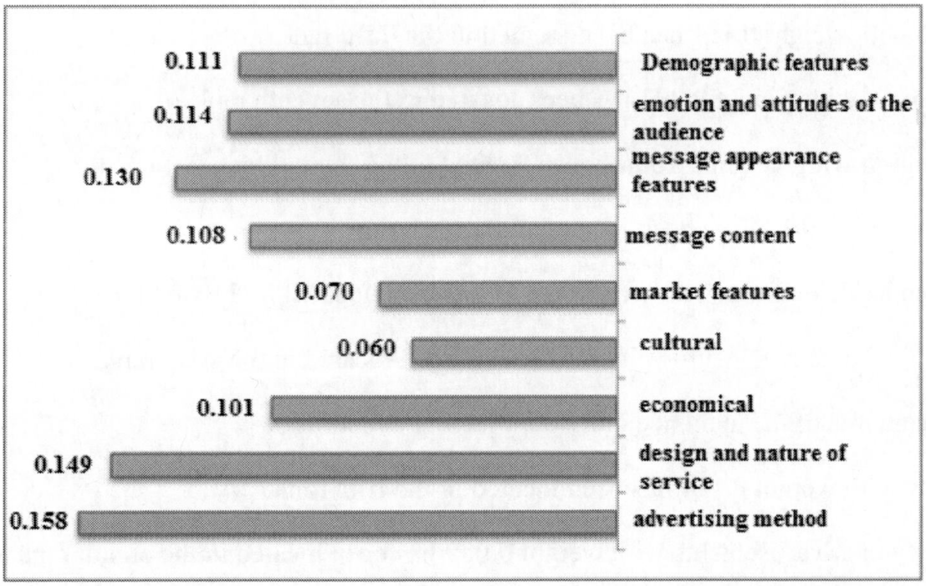

Figure 2- graphical display of significance of criteria of factors effective on advertisement effectiveness

Based on the obtained special vector:

- Demographic features with weight 0.111 have been located in rank five

- Individual feelings and attitude of the addressee with weight 0.114 has been located in rank four.

- The message surface features with weight 0.13 has been located in rank three.

- The message content with weight 0.108 has been located in rank six

- The market feature with weight 0.07 has been located in rank eight.

- Cultural with weight 0.06 has been located in rank nine.

- Economic with weight 0.10 has been located in rank seven.

- Plan and nature of services with weight 0.149 has been located in rank second.

- Advertisement methods with weight 0.158 has been located in rank first.

The inconsistency rate of performed comparisons has been obtained 0.030 which is smaller than 0.1 and so the performed comparisons is sufficed.

2- Comparison and determining priority of sub standards

At first, the experts' views have been collected in the dual comparison of sub standards. Then, the performed calculations for fuzzification of the experts views average have been provided for determining the priority of sub standards.

Priority of sub standards of demographic features

The audience job with weight 0.151 has been located in the third rank.

The audience income with weight 0.093 has been located in the seventh rank.

The audience religion with weight 0.16 has been located in the second rank.

Social class with weight 0.141 has been located in the fifth rank.

Awareness of gender differences with weight 0.173 has been located in the first rank.

The individual knowledge level with weight 0.135 has been located in the sixth rank.

Priority of sub standards of the audience individual feeling and attitude

Positive mentality with weight 0.124 has been located in the fifth rank.

Feeling of honesty in advertisements with weight 0.089 has been located in the eighth rank.

Threatening feeling with weight 0.14 has been located in the third rank.

Mental image with weight 0.102 has been located in the seventh rank.

Attitude towards brand with weight 0.121 has been located in the sixth rank.

Convincing audiences with weight 0.149 has been located in the first rank.

Attitude towards advertisement with weight 0.143 has been located in the second rank.

The individual perception of accuracy of advertisement with weight 0.133 has been located in the fourth rank.

Priority of sub standards of the message surface features

Then, based on the results of performed dual comparisons:

Beautiful and appropriate combination of colors with weight 0.143 has been located in the fourth rank.

Attention to reduction of the user waiting time with weight 0.159 has been located in the second rank.

Graphics- design with weight 0.131 has been located in the sixth rank.

Creativity in the message with weight 0.137 has been located in the fifth rank.

Emotional music with weight 0.099 has been located in the seventh rank.

Text repetition with weight 0.153 has been located in the third rank.

Access to required and sufficient information with weight 0.178 has been located in the first rank.

Determining the priority of sub standards of the message content

Benevolent with weight 0.093 has been located in the eighth rank.

Coordination in speech and behavior with weight 0.125 has been located in the fourth rank.

Reality of the messages with weight 0.119 has been located in the sixth rank.

Using women and children images with weight 0.131 has been located in the third rank.

Gender of speaker in the advertisement with weight 0.106 has been located in the seventh rank.

Using trustful and famous people with weight 0.16 has been located in the first rank.

Loveliness of famous people for the observer with weight 0.122 has been located in the fifth rank.

Familiarity of the famous person with the observer with weight 0.144 has been located in the second rank.

Priority of sub standards of the market features

Then, based on the results of the performed dual comparisons:

Factors relating to its application in the market with weight 0.326 have been located in the second rank.

Awareness of different aspects of status and environment with weight 0.283 has been located in the third rank.

Awareness of evaluation method of service receiver with weight 0.391 has been located in the first rank.

Priority of cultural sub standards

Then, based on the results of performed dual comparisons:

Awareness of norms with weight 0.199 has been located in the third rank.

Awareness of religion and religious values with weight 0.238 has been located in the first rank.

Ethical beliefs with weight 0.18 have been located in the fourth rank.

Obviating primary needs of the audiences with weight 0.164 has been located in the fifth rank.

Reference groups with weight 0.219 have been located in the second rank.

Determining the priority of economic sub standards

Then, based on the results of the performed dual comparison:

Economic beliefs with weight 0.261 have been located in the second rank.

Strategic look at an advertising program with weight 0.215 has been located in the fourth rank.

Market share with weight 0.274 has been located in the first rank.

Attention to economic security with weight 0.251 has been located in the third rank.

Priority of the plan and services nature sub standard

Awareness of services with weight 0.104 has been located in seventh rank.

Expressing potential benefits of purchase and services consumption with weight 0.117 has been located in the third rank.

Services complexity rate with weight 0.089 has been located in the ninth rank.

Uniqueness with weight 0.108 has been located in the sixth rank.

Rate of familiarity of the services consumer with weight 0.12 has been located in the second rank.

Price and allowances with weight 0.114 has been located in the fourth rank.

Relation with the customer with weight 0.111 has been located in the fifth rank.

Displaying the expected experience with weight 0.098 has been located in the eighth rank.

Supporting customers with weight 0.139 has been located in the first rank.

Priority of sub standards of the advertising methods

Beginning the advertisement with a challenge with weight 0.079 has been located in the ninth rank.

Beginning an advertisement with a statement and explanation with weight 0.107 has been located in the fifth rank.

Using stimuli with diversions more than usual and expected with weight 0.097 has been located in the seventh rank.

The advertisement message text being predicating with weight 0.126 has been located in the third rank.

Using diverse methods with weight 0.088 has been located in the eighth rank.

Media advertisement with weight 0.113 has been located in the fourth rank.

Dimensions of TV advertisement with weight 0.151 have been located in the first rank.

Psychological, geographical division with weight 0.107 has been located in the fifth rank.

Advertising media selection with weight 0.132 has been located in the second rank.

Determining the final priority of major criteria with FAHP technique

For determining final priority of major criteria using FAHP technique, the weights relating to major criteria (W1) and weight of indicators based on each criteria (W2) are needed. The results of comparison of the research sub criteria relating to them forms matrix W2. For determining the final priority of indicators with AHP technique, it is sufficient that the weight of indicators based on every criterion (W2) is multiplied in the major criteria (W1). Any of these matrices has been calculated in the previous steps. Based on the obtained output, it is obvious that the market share item, one of economic factors with weight 0. 277 is the most important indicator in the existing indicators. Item of awareness of evaluation method of service receiver in relation to other services with weight of 0.0273 has the second priority. Item of economic believes, one of economic factors, has the third priority.

So, the research model is as follows:

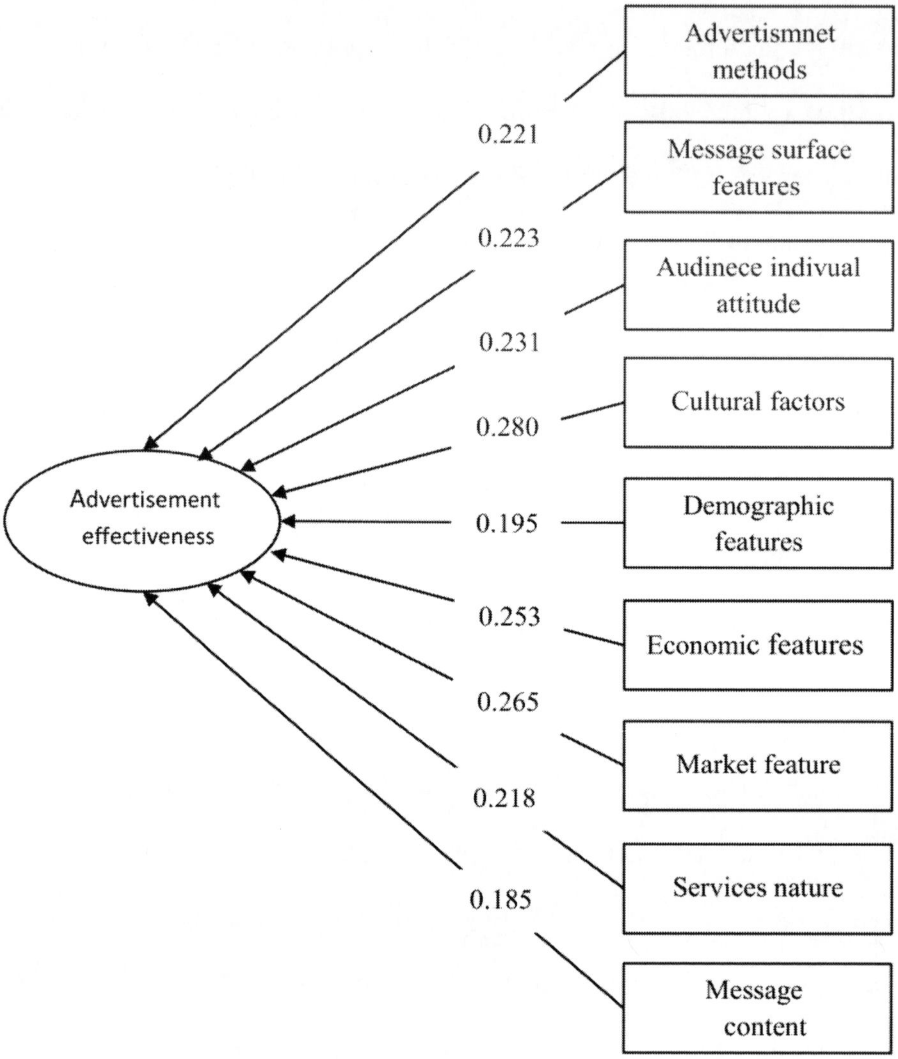

4- Conclusion:

In this study, identification and ranking of factors effective on designing the native model of advertisement in Eghtesad Novin bank has been addressed. In this first step, the primary indicators were identified and selected. In the first stage, by reviewing the literature and performed studies, a great number of indicators were extracted from the literature. For screening the indicators and identifying final indicators, fuzzy Delphi approach has been used. The experts' views have been collected in the field of significant rate of major indicators. From 84 indicators which were assessed, 24 indicators which have acquired score less than 7 are omitted. So, 60 indicators were finally confirmed. Then, regarding the performed calculations, the final weight of any of the model indicators have been calculated with fuzzy AHP. Based on the obtained output, it is obvious that the market share of economic factors with weight of 0.027 is the most important indicator in the existing indicators. The item of awareness of evaluating services in relation to other services is one of market features factors with weight 0.0273 which has second priority. The item of economic beliefs with weight 0.0264 is in the third priority.

Today, considering extension of competition in the international level and formation of world competition, all countries of the world may be successful in this arena to coordinate them with this trend and create more value for their customers by services and products they provide. In the today modern world, almost all people are exposed to advertisements and other communicational activities. Effectiveness of advertisement is one of important issues in the advertisement domain. In this regard, Mansouri Moyed (2005) showed that attention attraction, commercial advertisements of Maskan Bank are effective but in respect of other variables, that is, creating interest, stimulation of tendency and measurement to purchase, commercial advertisements of Maskan Bank are not effective. Samsam Shariat (2007) showed that TV channels and cartoon animations , attractive package, colors relating to the product and innovation, noticing safety feeling and associated symbols, informing of quality, background and price, dialects and common phrases, diversity in the message content, in TV, broadcasting from 18-23 and before beginning programs (in media) have been effective factors in effectiveness of commercial effectiveness from people viewpoint. Mosavi Esmaeilzadeh (2015) showed that oral advertisements (in 4 dimensions) and after that TV advertisement (just in 2 dimensions) are effective. Also, radio, internet and billboard advertisement are not effective. And also Friedman test was used for ranking advertisement tools. Salimi Fard & Pourdehghan (2016) showed that among factors effective on attitude to cell phone advertisement, components of general attitude towards advertisement, knowledge level and value have respectively the greatest significance and components of time of presentation and reference groups with the least significance in effecting the individuals attitude towards cell phone advertisement. Ansari & Rasaei (2016) showed that the creativity of advertisement message , market research, competition, market share, uniqueness and relation with customer are effective on the effectiveness of brand advertisement. Louis (2012) showed that advertising in the urban public spaces will have a great impact in the consumers' mentality. Since some advertisements are transitory and are often displayed in a short period of time and just in several spaces, but advertisement in public spaces exist everywhere and is similar to advertising through print and media, but since they have lower cost for production and distribution, have lower cost in this respect. Zabckowich & Loukston (2011) showed that in necktie advertisements use more of sensory requests and information signs. Findings showed that in western and European countries, people establish more contact due to globalization increase. They declared that content of printed advertisements, includes visual and text advertisements which are effective in establishing relation with many audiences. Also, ethnic identity as an important cultural variable increases attention to consumer purchase behavior. Besides that, they found out that the image of people, models, ceremonies and managers constitute an important part of images and background image plays an important role in effectiveness of visual factors in advertisement. Mac Ki- Nesbiat et.al (2011) showed that younger adults remind emotional messages specially negative ones better than others and older adults more prefer positive and logical messages to negative and emotional ones, but attitude towards advertisement among younger adults is not different. Also, they found that older adults have the most positive attitude towards advertisements.

Perzouti (2015) showed that advertising models are different in various ages and each age is consistent with a special model and using various and consistent models with people conditions may make the advertisement more effective. Abri & Joumar (2014) showed that advertising messages are effective on the consumer behavior and the advertiser may use various media for communicating consumer and media schedule is effective on the advertisement success. Attention to environmental factors where the customer lives are important factors in the advertisement success. Madra et.al (2017) showed that the advertisement text, text repetition impact the advertisement effectiveness. Maryion et.al (2017) showed that simultaneous use of several media like TV and internet may increase effectiveness of the advertisement and considering two variables of gender and multimedia create a better individual attitude towards the message effectiveness.

In ranking the identified factors, advertisement methods have assigned the first rank among factors. It is suggested that in designing advertisement and selecting advertising instruments, the dimensions of TV advertising, selecting median and predicative message text should be noticed. Using advertising instrument like internet advertisement, adverting on public vehicles like buses and metro due to great number of people who use them can increase stability of the advertisements in the customers mind.

In codifying and designing bank advertisements, the design and services nature should be noticed. It is proposed that supporting customer, familiarity of consumer with services are considered and during the advertisement process, it is required that the potential benefits of purchase and services consumption are expressed.

In designing the advertisement and content of bank advertising message, it is required to use famous and trustful people for introducing the services. Also, children as future customers and audiences should be noticed since they are future potentials. Total advertisements provided by the bank through various media should move in a unit line so that the audience can establish a logical relation between the observed advertisements in various media that is total advertising efforts lead to synergy of advertisement effectiveness. This issue won't be acquired unless through a media planning and efficient advertisement.

In advertisements, the message surface features should be noticed. It is required that in advertisement, access to required and enough information for the customers, attention to reduction of expectation time of user for observing the advertisement and suitable repetition of the text are considered. Providing large information in an advertisement which has a short time and providing information that require special mathematical analysis should be avoided.

References:

Hosseinabadi, Maryam, Mahmoud Nematian, Mohammad Besharati, investigating the relation between mixed elements of marketing and customers satisfaction with Tehran

Sarmayeh bank branches, management and entrepreneurship studies, fall 2016, 2nd period, No. 3 (page 162-169)

Dargi Parviz, 2008, subjects and topics of marketing management with Iran market attitude, Tehran, Rasa publication

Sarmad, Zohreh, Bazargan Abbas, Hejazi, Elaheh, 2014, research methods in behavioral sciences, Tehran, Agah. 27-13

Salimi Fard, Khodakaram, Pourdehghan, Adel, 2016, identifying and ranking factors effective on attitude towards cell phone advertisement using AHP, Brand management, period 3, no. 6, summer 2016, page 133-160

Seyed Akbar Manjgani, Vahideh Tabasi Lotfabadi, 2015, factors effective on effectiveness of advertisement on financial stoop of legal clients of Mashhad City with AHP approach, finance journal, fall 2015, No. 75 (page 121-146)

Sanayei, Ali, Majid Mohammad Shafiei, Maedeh Amini Valashani, 2016, effectiveness of bank advertisement in customer attraction, commercial management searches, spring and summer 2016, No. 15 (page 185-210)

Mahmoudian, Mahmoud, Rabiei, Ali, Jamili, Bita. 2011, evaluation of effectiveness of Parsian bank advertisement and identifying the most important factors in increasing its effectiveness in Tehran city, modern marketing studies, summer 2011, No. 2 (page 17-40).

Manjagani, Seyed Akbar, Tabasi Lotfabadi, Vahideh, 2015, factors effective on advertisement effectiveness on financial stoop of legal clinets of Mashhad city with AHP approach, finance journal, fall 2015, No. 75, 1-121

Venous, Davar, Safaeian, Mitra, (2005), marketing bank services in Iran and bank services features, Tehran, Negah Danesh, 4th edition

Angulo, N. (2016). 10 digital marketing trends to watch in 2016 and beyond. Retrieved from: http://www.marketingdive.com/news/10-digital-marketing-trends-to-watch-in-2016-./and-beyond/411457

Bell Raoul, Buchner, Axel.(2018). Positive Effects of Disruptive Advertising on Consumer Preferences, Journal of Interactive Marketing 41 (2018) 1–13

Corvi. Elisabetta.(2010).The effectiveness of advertising: a literature review, 10th Global Conference on Business & Economics, ISBN : 978-0-9830452-1-2

Cramphorn Michael F.,(2011), Forum Gender effects in advertising, International Journal of .Market Research Society Vol. 53 Issue 2

Cronin, J. J. & Fox , G. L. (2010). "The implications of third party customer complaining for .advertising efforts". Journal of advertising. 39(2): 21-23

Decrop, A. (2007). The influence of message format on the effectiveness of print advertisements for tourism destinations. International Journal of Advertising, 26(4), 505–.525

Denis .A. Shakhov and ,Anna A. Panasenko.(2012).Evaluating Effectiveness of Bank Advertising in the Internet: Theory and Practice, World Applied Sciences Journal 18 (Special Issue of Economics): 83-90, 2012 ISSN 1818-4952

Denis 1,2 A. Shakhov and 3Anna A. Panasenko.(2012).Evaluating Effectiveness of Bank Advertising in the Internet: Theory and Practice, World Applied Sciences Journal 18 (Special Issue of Economics): 83-90, 2012

Durdana Ozretic-Dosen Ines Zizak , (2015),"Measuring the quality of banking services targeting student population", EuroMed Journal of Business, Vol. 10 Iss 1 pp. 98 - 117

Eckler, P., and Bolls, P. (2011). Spreading the Virus: Emotional Tone of Viral Advertising and Its Effect on Forwarding Intention and Attitudes. Journal of Interactive Advertising, .11(2): 1-11

Eric W.T. Ngai Ka-leung Karen Moon S.S. Lam Eric S. K. Chin Spencer S.C. Tao , (2015),"Social media models, technologies, and applications", Industrial Management & Data Systems, Vol. 115 Iss 5 pp. 769 – 802

Fotini Patsioura Maro Vlachopoulou Vicky Manthou, (2009),"A new advertising effectiveness model for corporate advertising web sites", Benchmarking: An International Journal, Vol. 16 Iss 3 pp. 372 - 386

Frederic Crane, , G. (2010). «Marketing FO Entrepreneurs: ConceptandApplication for new .Venture. New York: SAGE Pablication, Inc

Frolova. Svetlana.(2014).THE ROLE OF ADVERTISING IN PROMOTING A PRODUCT, Thesis CENTRIA UNIVERSITY OF APPLIED SCIENCES ,Degree Programme in Industrial Management May 2014, p"1-61

Henry Boateng Abednego Feehi Okoe , (2015),"Consumers' attitude towards social media advertising and their behavioural response", Journal of Research in Interactive Marketing, Vol. 9 Iss 4 pp. 299 -312

Henry Boateng Abednego Feehi Okoe , (2015),"Consumers' attitude towards social media advertising and their behavioural response", Journal of Research in Interactive Marketing, Vol. 9 Iss 4 pp. 299 -312

Huang. Jinsong, Song Sub, , Liuning Zhouc, , , Xi Liu .(2013).Attitude Toward the Viral Ad: Expanding Traditional Advertising Models to Interactive Advertising, Journal of Interactive Marketing, Volume 27, Issue 1, February 2013, Pages 36–46

Idris, I., M.S.A. Yajid and A. Khatibi, 2009. Personal and Psychological Factors-Does it .Impact the Choice of Advertising Medium? Journal of Social Sciences, 5(2): 104-111

Ioannou Myria, Nikolaos Boukas, Evi Skoufari .(2014).Examining the role of advertising on the behaviour of co-operative bank consumers, Journal of Co-operative Organization and Management, Volume 2, Issue 1, June 2014, Pages 24-33

Jankowski. Jarosław, Przemysław Kazienko, Jarosław Wątróbski, Anna Lewandowska, Magdalena Zioło.(2016).Fuzzy multi-objective modeling of effectiveness and user experience in online advertising, Expert Systems with Applications, Volume 65, 15 December 2016, Pages 315-331

Kim, C., Park, S., Kwon, K., & Chang, W. (2012). An empirical test to measure the effectiveness of online advertising in online marketplaces using a hierarchical Bayes model. .Expert Systems with Applications (39), 117–128

Kim, ji-Hern., Hyun, yong j., (2010). " A model to investigate the influence of marketing – mix efforts and corporate image on brand equity in the IT software sector " . industrial . Marketing Management. No 40,2011, pp 424-438, Emerald Group Publishing Limited

Kim, S., & Jun, J. (2016). The impact of event advertising on attitudes and visit intentions. .Journal of Hospitality and Tourism Management, 29, 1-8

Kirby, J. and Marsden .(2006). Connected marketing : the viral, buzz and word of mouth .revolution, Oxford : Butterworth-Heinemann, 2006

Madera. Quetzali, Oscar Castillo ∗, Mario García-Valdez, Alejandra Mancilla..(2017). A method based on Interactive Evolutionary Computation and fuzzy logic for increasing the effectiveness of advertising campaigns, Information Sciences 414 (2017) 175–186

Marion Garaus and Udo Wagner , Anna-Maria B¨ack.(2017). The Effect of Media Multitasking on Advertising Message Effectiveness, Psychology & Marketing, Vol. 34(2):)138–156 (February 2017

Martín-Santana J.D., E. Reinares-Lara , P. Reinares-Lara .(2017).Influence of radio spokesperson gender and vocal pitch on advertising effectiveness: The role of listener gender, Spanish Journal of Marketing – ESIC, Volume 21, Issue 1, February 2017, Pages 63-71

McKay-Nesbitt, J., V. Manchanda, R., C. Smith, M., & A. Huhmann, B. (2011). Effects of age, need for cognition, and affective intensity on advertising effectiveness. Journal of Business Research (64), 12–17

Moch, Nils (2013). Competition in fragmented markets: New evidence from the German banking industry in the light of the subprime crisis. Journal of Banking &; Finance, Volume 37, Issue 8, Pp 2908–2919

Moorthy, S. and Hawkins, S. (2005). Advertising repetition and quality perception. Journal of .Business Research, 58, 354– 360

Mucan. Burcu, Özeltürkay. Eda Yaşa.(2014).Social Media Creates Competitive Advantages: How Turkish Banks Use This Power? A Content Analysis of Turkish Banks through Their Webpages, Procedia - Social and Behavioral Sciences, Volume 148, 25 August 2014, Pages 137-145

Muda, M.; Musa, R.; Niana-Mohamed, R. and Borhan, H. (2014). Celebrity Entrepreneur .Endorsement and Advertising Effectiveness. Social and Behavioral Sciences, 130, 11-20

Oladele. Kehinde Joseph, Ogunnaike, Akinbode Mosunmola, Aka Deborah .(2016). Effective Advertising: Tool For Achieving Client-Customer Relationships, Researchjournali's Journal of Media Studies Vol. 2 | No. 1 February | 2016

Olsen, G., Pracejus, J. W., & O'Guinn, T. (2012). Print advertising: White space. Journal of .Business Research (65), 855–860

Park, M. (2006). The compensatory effects of pictorial and verbal information for haptic information on consumer responses in non-store shopping environments. Unpublished Ph.D, .Ohio, Ohio state University

Park, T. (2008). Effective design of advertisement in handheld devices (Doctoral dissertation .)PURDUE UNIVERSITY

Park, T. (2008). Effectiveness Design of Advertisement in Handheld Devices. Dissertation for .doctoral degree, Purdue University, West Lafyette India

Pergelova, A., D. Prior and J. Rialp, 2010 Assessing Advertising Efficiency Does the Internet Play a Role? Journal of Advertising, 39(3): 39-54

Pilelienė Lina Viktorija Grigaliūnaitė , (2017)," The effect of female celebrity spokesperson in FMCG advertising: neuromarketing approach ", Journal of Consumer Marketing, Vol. 34 - .Iss 3 pp

Rosemary J. Averya, Matthew D. Eisenbergb, Kosali I. Simon .(2012).The impact of direct-to-consumer television and magazine advertising on antidepressant use, Journal of Health Economics, Volume 31, Issue 5, September 2012, Pages 705–718

Selviaridis. Kostas, Martin Spring b, Luis Araujo b .(2013). Provider involvement in business service definition: A typology, Industrial Marketing Management 42 (2013) 1398–1410

Strebinger .Andreas, Xiaoling Guo, Ferdinand Klauser, Peter Grant-Hayd .(2018).Is Multi-Ethnic Advertising a globally viable strategy for a Western luxury car brand? A mixed-method cross-cultural study, Journal of Business Research 82 (2018) 409–416

Suvittawat, A., (2015). "Commercial Bank Service Quality: Chonburi Province, Thailand". .International Journal of Management Sciences and Business Research. 4(2): 2226-8235

Tafani. Eric, Elyette Rouxa, Rainer Greifeneder .(2018).In the mood for action: When negative program-induced mood improves the behavioral effectiveness of TV commercials, Journal of Business Research 84 (2018) 125–140

Te'eni-Harari, T. (2014). "Clarifying the relationship between involvement variables and .advertising effectiveness among young people". Journal of Consum Policy. 37: 183–203

Thomas, S. and Kohli, S.C. (2011). Can Brand Image Move Upwards after Sideways? A .Strategic Approach to Brand Placements. Business Horizons, 54 (1): 41-49

Yahyaoui . Hamdi, Own .Hala S..(2018). Unsupervised clustering of service performance behaviors, Information Sciences 422 (2018) 558–571

Yuksel, A., Yuksel, F. & Bilim, Y. (2010) "Destination Attachment: Effects on Customer Satisfaction and Cognitive, Affective and Cognitive Loyalty", Journal Of Tourism Management, Vol.31, pp: 274-284.

STUDY OF NATURAL ANTIOXIDANT INHIBITORS FOR CORROSION PROTECTION OF MILD STEEL IN ACIDIC MEDIUM

Ms. Ibtihal Hassan AL-Bahrani, National University of Science and Technology, Sultanate of Oman.

ABSTRACT

Corrosion is a natural phenomena and it is a very big loss for structure as well as business; so corrosion minimization is one of the biggest areas of research. In this project; corrosion inhibition efficiency of berries juice extract was tested in mild steel sample against the sulfuric acid medium which was prepared from four types of berries; blackberries, blueberries, cranberries, and raspberries.

The experiment was designed by using Design Expert V 11 where immersed time, acid and berries juice extract concentrations were set as factors and a limited range was selected. The corrosion rate was set as a response. 20 runs were under taken and the experiments were done in the lab with maximum care.

The corrosion rate of 1020.374 mpy minimized to 597.659 mpy in presence of 5ml of extract and the efficiency of berries juice extract reached to 41.427%. The RSM gave 232.022 mpy corrosion rate as a minimum value and the optimum equation given by RSM was used for Genetic Algorithm analysis which gave 188.866 mpy corrosion rate at the optimizing conditions.

The surface and chemical characterization of the mild steel sample in presence and absence of berries juice extract were analyzed along with the relations between temperature, immersed time, acid and berries juice extract concentrations with response. The mixing tank was designed, calculations and drawing were presented.

KEYWORDS

Corrosion, Mild steel, Berries, Inhibitor, Weight loss, Sulfuric acid, RSM, SEM.

INTRODUCTION

Mild steel is used in most of our daily life applications as well as in industrial applications. In other hands, mild steel has low corrosion resistance which corrodes easily once it reacts with the surrounding environment; corrosion is a phenomenon where the mild steel reacts with the deteriorates and surrounding environment. Corrosion of mild steel is a continuous and constant problem where it severe financial and life loses as it leads to loss in the product, efficiency of equipment and failures in some cases (Roberge,2012).

The corrosion phenomenon controlled by adding a small amount of component to the corrosive environment which is a corrosion inhibitor. There are organic and inorganic inhibitors. The inorganic inhibitors are used in limit amount as they are toxic such as nitrites, molybate, nitrate, silicate, chromate, and phosphate (Cafferty,2010). Now a day, the corrosion engineers looking for cheap, renewable, friendly, readily available and ecologically acceptable inhibitors which are organic inhibitors which comes from the plants such as antioxidant components consisting of tannins, saponins, flavonoids, alkaloids and amino acids. Organic inhibitors consist of some heteroatoms such as oxygen, nitrogen, multi-bound, and sulphur atoms with the conjugated system which allows the absorption in the metal surface. Those atoms attract the absorption of some components on the metal surface and that done by the interaction of metals and by the electron of the conjugated system or by forming a bond with the free electron of heteroatom which forms a layer in the metal surface, as a result, it decreases the corrosion rate (Roberge,2012).

Berries are small pulpy fruits which have up to 9.2 mmol of antioxidant per 100g of berries. Berries can be used as a corrosion inhibitor as they are rich in antioxidant. The studies show that the berries juice extract of blueberries, cranberries, raspberries, and blackberries rich with antioxidant where the antioxidant capacity of them reach to 10 mg/ 100 g of blueberries (Michalska&Lysiak,2015), 58 mg/100g of cranberries (Blumberg,J.,et al,2013), 26.2mg/ 100 g of raspberries (Bobinaite,R.,Viskelis&Venskutonis,2015), and 21 mg/100 g of blackberries (Kaume,L.,et al,2012).

OBJECTIVES

To prepare anticorrosive extract (berries), to study corrosion in sample in presences and absences of the inhibitors, to determine the optimum conditions to inhibit corrosion and to design mixing tank.

METHODOLOGY, DESIGN AND EXPERIMENTATION

PREPARATION OF MILD STEEL SAMPLES

The mild steel sheets were cut into 2.5*3 cm^2 by using steel cutter and cleaned by surface cleaning machine to remove any corroded products in the metal surface then 4mm hole done by using drilling machine. Bunches for all metal samples (1-20) were made to differentiate between them. The drilled samples were washed by distilled water then rinsed with benzene and acetone respectively and dried with a hot air dryer. They were weighted using sensitive balance and kept in the desiccator until the time of usage.

PREPARATION OF CORROSIVE MEDIUM

The corrosive medium of different concentrations of H_2SO_4 prepared by using 36.8 N of the sulfuric acid.

PREPARATION OF BERRIES JUICE EXTRACT

137g, 209g, 282g, and 200g of raspberries, cranberries, blueberries, and blackberries respectively have been used to produce 309.5 ml of berries juice extract. The different types of berries washed by cold water, crashed in the electrical crasher and compressed mechanically by hand then filtered by cloth filter. The filtrate filtered again by using filter paper. The berries juice extract kept in the freezer at 0°C till the time of usage.

The efficiency of berries juice extract has been tested by running three experiments. The cleaned and weighted mild steel samples immersed in 0ml(blank), 2ml and 5ml of berries juice extract respectively with 50 ml of 1M H_2SO_4. After 2 hours; the samples have been taken out and cleaned by distilled water, benzene and acetone then dried by air dryer and weighted again. The corrosion rate and inhibitor concentration have been calculated by using equations (1) and (2) respectively.

$$Cr(mpy) = 22300\times\frac{\Delta W}{\rho\times A\times t} \quad \textbf{equation(1)}$$

$$IE(\%) = \frac{\text{initalcorrosionrate-finalcorrosionrate}}{\text{initalcorrosionrate}}\times 100 \quad \textbf{equation(2)}$$

DESIGN THE EXPERIMENT CONDITIONS

The experiments conditions have been designed by using Design Expert software-version 11 and following Response Surface Module(RSM); the range of the parameters have been given as inputs to the software which are inhibitor concentration (5-15 ml), immersed time (180-240 min) and acid concentration (1-2 M) and the response set as corrosion rate in mpy. Table(1) shows the experiment conditions given by the software.

Table(1): Designed runs given by RSM.

Run.No	Factor(A): inhibitor, ml	Factor(B): Time, min	Factor(C): Acid concentration, M
1	10	180	1
2	15	240	2
3	5	120	1
4	5	240	1
5	10	180	1.5
6	15	180	2
7	10	120	1.5
8	5	120	1.5
9	10	180	2
10	10	180	1.5
11	10	180	1.5
12	15	240	1
13	15	120	1
14	5	240	2
15	10	180	1.5
16	15	180	1.5
17	5	120	2
18	10	180	1.5
19	10	180	1.5
20	10	240	1.5

APPLYING THE DESIGNED RUNS IN THE LABORATORY

The weighted mild steel sample has been hanged by using a plastic thread in a 50 ml of sulfuric acid in presence of berries juice extract of known concentration and contact time as per the software; the batch has been kept in the shaker. After the contact time the mild steel sample cleaned by using 120 grade emery paper to remove any corroded products, washed by distilled water, rinsed by benzene and acetone respectively and dried by hot air dryer, weighted by using sensitive balance. The corrosion rate in mpy was calculated by using equation(1) where the weight losses results have been used.

STUDYING THE IMPACT OF TEMPERATURE IN CORROSION RATE

To study the impact of temperature in the corrosion rate; mild steel samples weighted and hanged for 2 hours in 50 ml of 1M sulfuric acid with 10ml of berries extract at different temperatures(30, 40, 50 and 60 °C) as shown in Table(2); each batch has been kept in the water bath shaker with heater to provide perfect contact between the solution and mild steel sample and to maintain the required temperature. Corrosion rate has been calculated by equation(1).

Table(2): Experiment conditions of temperature into corrosion rate.

SI.NO	Temperature (°C)	Inhibitor concentration (ml)	Acid concentration (M)	immersed time, (hours)
1	30	10	1	2
2	40	10	1	2
3	50	10	1	2
4	60	10	1	2

SURFACE AND CHEMICAL CHARACTERIZATION ANALYSIS

The mild steel surface of blank, absence and presence of berries juice extract samples have been studied by using Scanning Electron Microscopy (SEM) and Energy Dispersive Spectroscopy (EDS).

OPTIMIZATION STUDY BY GENETIC ALGORITHM

Matlab has been used for optimization where the genetic algorithm method has been followed. The actual equation form Design expert(RSM) has been used where "Optimtool" followed. The same limit of the conditions from RSM has been selected. The same previous procedure has been followed in the lab with maximum care by using those optimum conditions which given by the software, 1.001M of H_2SO_4 with 15 ml of berries juice extract immersed for 237.208 minutes.

DESIGN OF MIXING TANK

Design of mixing tank is a part of the project where the industrial scale mixing tank has been designed, the tank will be formed from high resistance material such as S.S to be able to resist the corrosive medium of sulfuric acid which will be mixed with berries juice extract. It is a container which used to mix various material (heterogeneous materials) together to form homogenous materials even the different viscosity materials can be mixed easily which provides continuous process, good control, easy cleaning, and simplest construction. It can be made from S.S material which provides a smooth surface and it can be easily cleaned (Petropedia,2018).

Parameters to be considered to design mixing tank are length, torque, weight, stiffness and hydraulic loading.(Dynamixianc,2015). The important factors in design mixing tank are tank type, tank volume, specific gravity, process and environment.(Ibryamoval,Y.et al,2009)

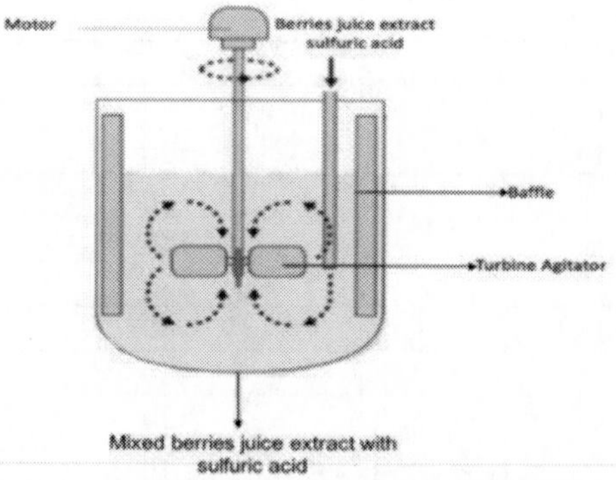

Figure(1): Mixing Tank(*Mixing Tank,*2018)

From the graph in Figure(2) Turbine agitator has been selected as it is suitable with the viscosity of the fluid and the tank volume. (Sinnot&Towler,2009)

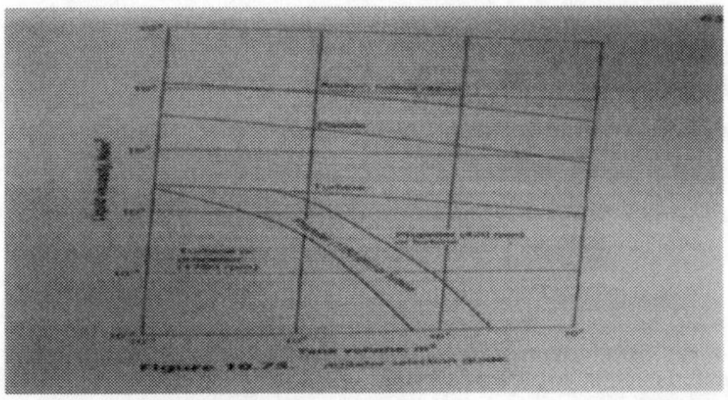

Figure(2): Graph of liquid viscosity Vs tank volume. (Sinnot&Towler,2009)

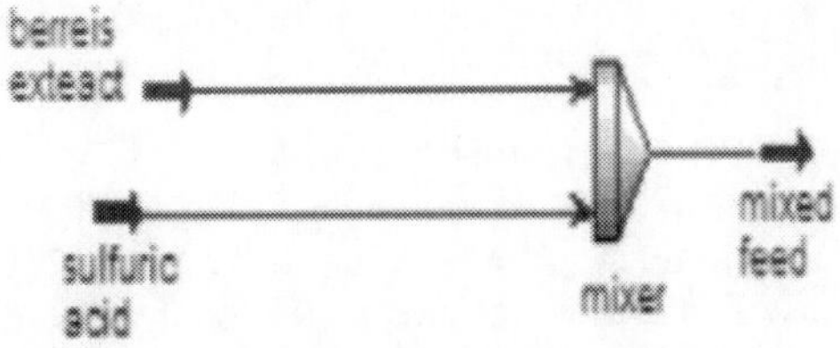

Figure(3): Result of Mixing Tank by Aspen Hysys.

Table(3): Calculated parameters for designing mixing tank

Tank diameter	m	3.5
Tank Height	m	1.81
Across sectional area	m^2	9.62
Volume of the tank	m^3	17.5
Impeller diameter	m	1.16
Length of paddle	m	0.29
Width of paddle	m	0.23
power	watt	204085
Reynolds Number	-	136486.6391

RESULTS AND ANALYSIS

EFFECT OF BERRIES JUICE EXTRACT CONCENTRATION INTO CORROSION RATE AND INHIBITOR EFFICIENCY

As shown in Table(4) the corrosion rate in the absence of berries juice extract inhibitor reach to the maximum (1020.374mpy); the corrosion rate has been minimized to 898.117mpy and 597.659mpy in presence of 2 and 5 ml of berries juice extract inhibitor respectively which means that berries juice extract inhibits the corrosion as the concentration of the berries juice extract increases. As well as the inhibitor efficiency reaches to 41.427% which obtained at 5ml of berries juice extract inhibitor in 1M of sulfuric acid immersed for 2 hours. The adsorption of the organic compounds such as ascorbic acid that present in the berries juice extract is blocking the active sites on the mild steel samples as it is forming a protective film on the metal surface as what kumar.A,2012 mention in his studies so the corrosion rate keeps decreasing as the amount of the berries juice extract is increasing. (Kumar.A,2012) Those compounds have heteroatoms such as oxygen which is a center of adsorption. In addition; It forms a coordinate bond as it has the ability to transfer the electrons to the vacant d-orbital of mild steel from the heteroatoms. (Ostovari, L. et al,2009)

Table(4): Experimental results of testing the efficiency of berries juice extract.

SI.NO	Inhibitor concentration (ml)	Corrosion rate (mpy)	Inhibitor efficiency (%)
1	0	1020.374	-
2	2	898.117	11.976
3	5	597.659	41.427

EFFECT OF IMMERSED TIME INTO CORROSION RATE AND INHIBITOR EFFICIENCY

The maximum corrosion rate reaches to 765.005mpy which obtained at 5ml of berries juice extract in the presence of 50 ml of 2 M of H_2SO_4 at 2 hours. Since the minimum corrosion rate of 232.022mpy obtained in presence of 5ml of berries juice extract with 50 ml of 1 M of H_2SO_4 immersed for 4 hours. The effect of immersed time into corrosion rate can be observed at low acid concentration and in the presence of berries juice extract, the corrosion rate is minimized at high immersed time. The decrease in the corrosion rate with the increase in immersed time is related to the adsorption of the corrosion product ($FeSO_4$) as shown in equation(3); According to Xianghong,L. et al ; the corrosion product of $FeSO_4$ becomes thicker with the immersed time, so it form a protective film (barrier) in the surface of mild steel (Xianghong, L. et al,2009).

$$Fe_{(s)}+H_2SO_{4(aq)} \rightarrow FeSO_{4(aq)} + H_{2(g)}\ ,\ \textbf{equation(3)}$$

EFFECT OF ACID CONCENTRATION INTO CORROSION RATE

The effect of acid concentration into corrosion rate has been studied by using 1, 1.5 and 2 M of H_2SO_4 at different inhibitor concentrations and immersed time where the corrosion rate increase with the acid concentration at low inhibitor concentration and immersed time which reach to 765.005mpy at 2M H_2SO_4, 5ml of berries juice extract which immersed for 120 minutes as a maximum corrosion rate as shown in Table (3). In presence of the corrosive medium of H_2SO_4 with less amount of berries juice extract; the mild steel surface was exposed to H_2SO_4 medium where the corrosion reaction started as shown in equation(3) and $FeSO_4$ was formed with H_2 bubbles.

EFFECT OF TEMPERATURE INTO CORROSION RATE

From Table(5) can be noted that the rate of corrosion increases with temperature where it reaches to maximum 2407.994mpy at maximum temperature of 60°C. According to Ambrish, S. et al, at high temperature the berries juice extract molecules were partially absorbed in the mild steel surface so it was exposed to the corrosive medium of H_2SO_4, as a result, increasing the corrosion rate (Ambrish,S. et al,2012).

Table(5): Results of studying the effect of temperature into corrosion rate

SI.NO	Temperature,° C	Corrosion rate, mpy
1	30	701.507
2	40	1666.404
3	50	2008.861
4	60	2407.994

ANALYSIS BY RSM

Once the experimental results and corrosion rate calculated the RSM analysis has been done where the corrosion rate has been given to the software as a response. The response must be modified where the results fit with linear. The fit summary shows that the model F-value and P-value of 12.66 and 0.0001 respectively imply that the model is significant where A, B, C, and B^2 are the significant terms in this model only. 1.48 is a lack of fit P-value which indicates that the fit is not significant as there are pure errors.

0.5562 of the predicted R^2 is a reasonable agreement with 0.7105 of the adjusted R^2. The signal to noise ratio of 13.7252 of Adeq precision that the ratio grater than 4 is desirable. (Mycourses,2016)

The process variables are studied using ANOVA. The estimate coefficients in term of coded factors are A-Inhibitor, B-time, C-Acid and B^2. It is creating an accompanying condition for the sample with 1.0000 correlation coefficient. The coded and actual equations given by software are shown in equation(4) and (5) respectively.

- Corrosion rate(mpy)= 370.52-40.52(A)-126.97(B)+ 80.85(C) +66.58$(B)^2$ **equation(4)**
- Corrosion rate(mpy)= 1189.15487-8.10479(Inhibitor) -8.77421(Time) +161.69302 (Acid Conc)+0.018495 $(\text{Time})^2$ **equation(5)**

The various relations can be studied from one factor graph; it can be noted that the corrosion rate increases with acid concentration and decreases with the increase in the amount of berries juice extract and immersed time as can be seen from the plots in Figure(3):Plot pf corrsoion rate Vs inhibitor concntration, Figure(4): plot of corrosion rate Vs acid concentration and Figure(5):Plot of corrosion rate Vs immersed time respectively.

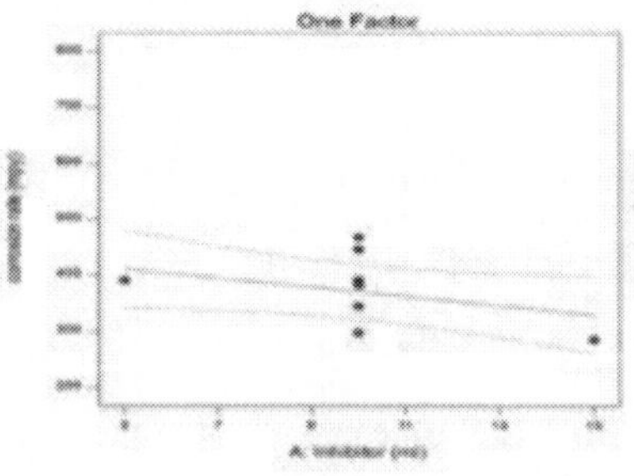

Figure(3)

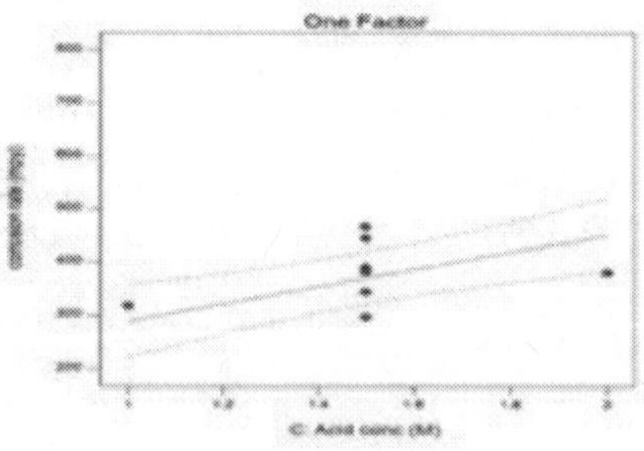

Figure(4)

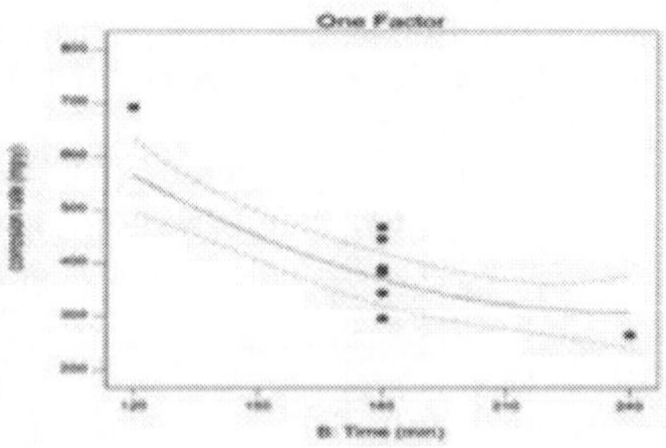

Figure(5)

The effects of two factors into the corrosion rate can be studied from the 3D plots as shown in Figure(6): 3D plot of corrosion rate Vs inhibitor concntration and acid concntration and Figure(7): 3D plot of corrosion rate Vs inhbitor concntration and immersed time.

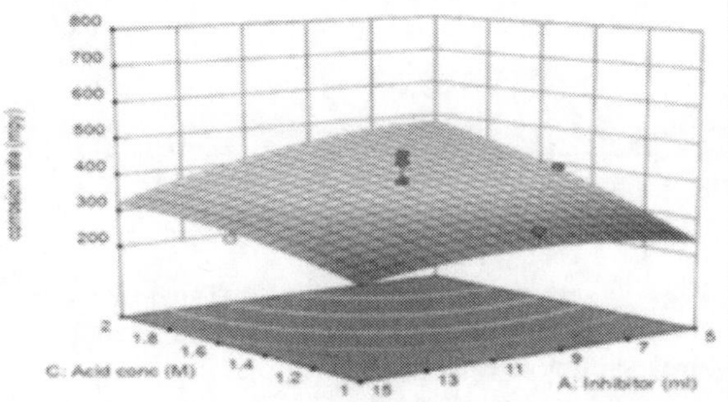

Figure(6)

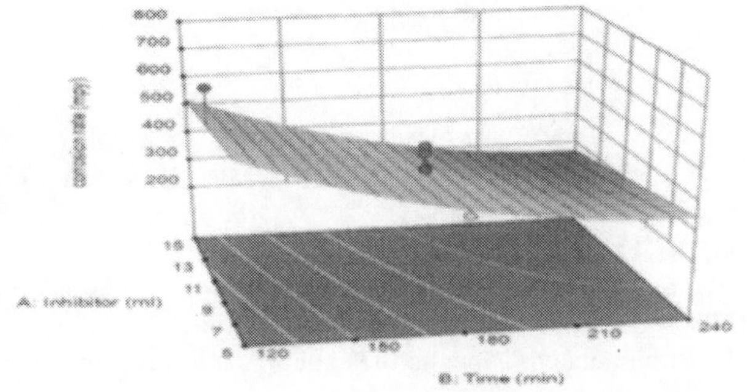

Figure(7)

The optimum conditions given by the RSM of 1M of H_2SO_4 with 5ml of berries juice extract inhibitor runs for 240 minutes gives 213.402mpy as a minimum corrosion rate but the experimental results at the same conditions gives 232.022mpy which means that they are too close as the suggested solution considered as an excellent optimum conditions.

SEM ANALYSIS

The SEM has been done for the blank mild steel sample for facilitating comparison as shown in Figure(8) which shows smooth mild steel surface compare to others. Rough surface in absence of berries juice extract is shown in Figure(9); where corrosion starting with the formation of H_2 bubbles as per the equation(3) which means that the surface of mild steel is exposed to the H_2SO_4 medium. In presence of 15 ml of berries juice extract, the surface of mild steel is protected by the adsorption of the organic compounds that are present in the berries juice extract where fewer cracks and pits can be noted and smooth surface is shown in Figure (10), which means that the mild steel surface is protected. (Raja, P. et al, 2008)

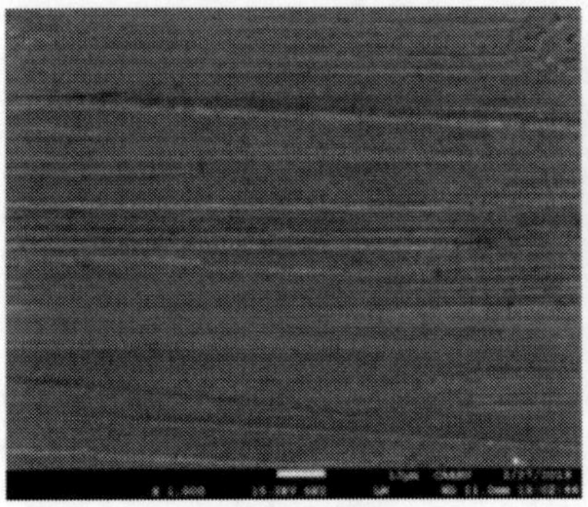

Figure(8)

Figure(9)

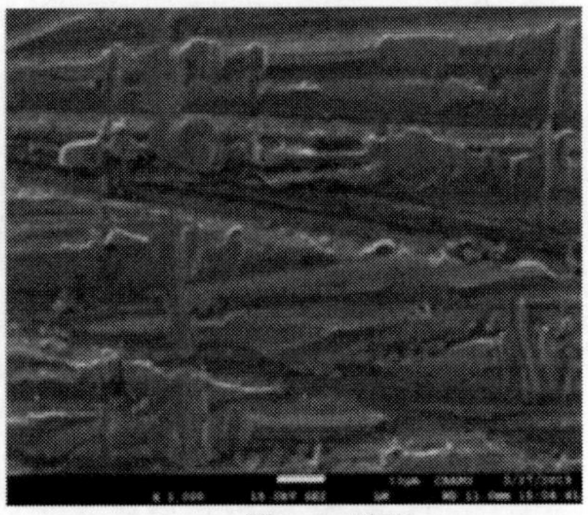

Figure(10)

EDS ANALYSIS

EDS has been used to study the chemical characterization or elemental analysis; Figure (11) shows the blank mild steel sample before being immersed in 1M H_2SO_4 in absence of berries juice extract where high Fe peak is shown, which indicates the presence of high amount of iron which reaches to 88 Wt% in the mild steel surface. Figure (12) shows the mild steel surface that has been immersed for 2 hours in 50 ml of 1M H_2SO_4 in absence of berries juice extract; the mild steel surface contains low amount of Fe (87.3 Wt%) compared to the blank mild steel sample. In Figure (13) the mild steel sample immersed for 2 hours in 50 ml of 1M H_2SO_4 in the presence of 5ml of berries juice extract, it shows high spark of Fe where the mild steel surface contains lower amount of Fe (84.8 Wt%) compared to Figure(11) and (12) which indicate lower iron dissolution rate in 1M H_2SO_4 medium containing berries juice extract (Alaneme, K. et al,2016). A high amount of oxygen and carbon are observed in presence of berries juice extract compared to the absence of berries juice extract. Si is shown in the absence and presence of berries juice extract which means that's it comes out to the surface during the reaction with sulfuric acid.

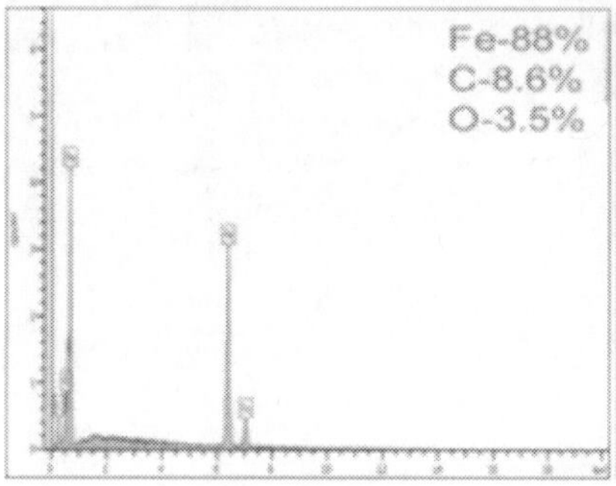

Figure(11)

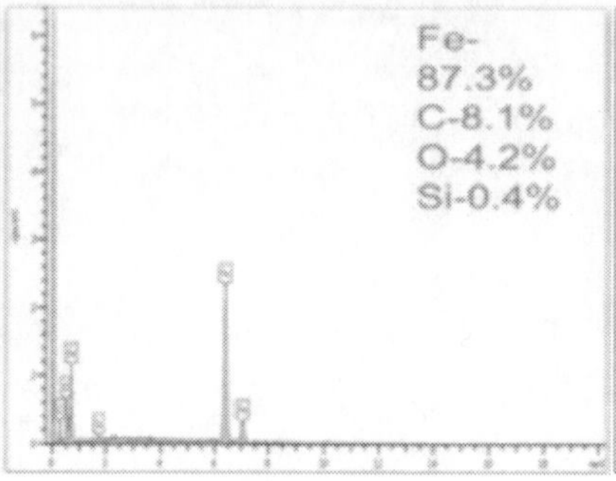

Figure(12)

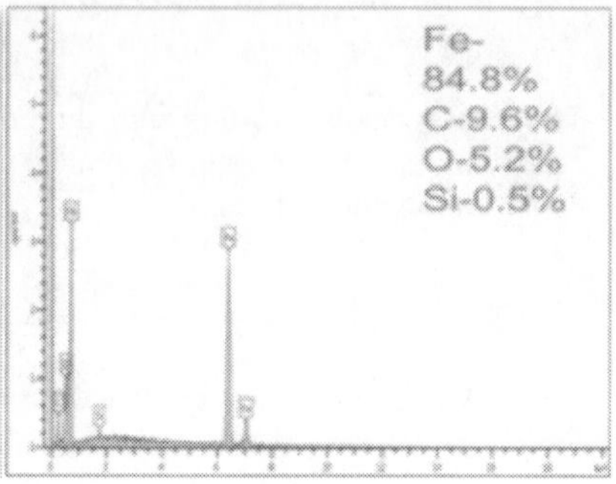

Figure(13)

GENETIC ALGORITHM OPTIMIZATION

The experimental result gives 288.367mpy as a corrosion rate. The software gives the minimum corrosion rate value of 188.866mpy from 50 runs as shown in Figure(14). A small difference can be noted as the genetic algorithm is not the exact fit with linear equations so the experimental and software results are too close to each other.

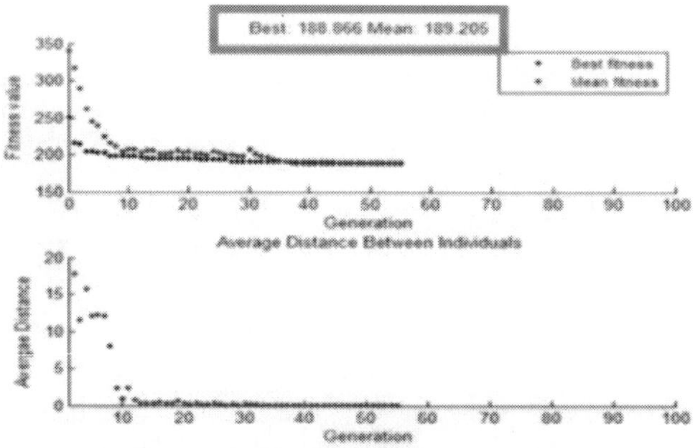

Figure(14): Genetic Algorithm optimization.

CONCLUSION

In this project, all the objectives have been achieved by designing and experimenting.

Firstly; the anticorrosive extract (berries) has been prepared, the prepared extract which contains 4 types of berries namely blueberries, blackberries, cranberries, and raspberries inhibit the mild steel corrosion in sulfuric acid medium with 41.427% efficiency in presence of 5 ml of berries extract.

Secondly, the mild steel corrosion in the presence and absence of berries juice extract were studied where the corrosion rate reaches to 1020.374mpy in absence of berries juice. By using 5 ml of berries juice extract the corrosion rate minimized to 232.022mpy.

Thirdly, the optimum conditions to inhibit the mild steel corrosion were studied by Design Expert V11. At the optimum conditions of 1M of H_2SO_4. 5ml of berries juice extract, 240 minutes and at room temperature the corrosion rate reaches to 232.022mpy as a minimum value.

The actual equation obtained from Design Expert V11 was later optimized using Genetic Algorithm. Genetic Algorithm shows that at 1.001M of H_2SO_4, 15 ml of berries juice extract and 237.208 minutes at room temperature, the corrosion rate is minimized to 188.866mpy.

Finally, mixing tank has been designed with 1.18m height and 2.5m diameter and a turbine agitator has been selected.

ACKNOWLEDGEMENT

I would like to thank god for giving me patience and power to work in my most important part of my life, my father, mother, my brother and my friends for their help and support. I would also like to express my sincere gratitude to Dr. Priy Brat Dwivedi for his support and help in developing my project idea, Mr. M.J.Varghese for his help in RSM analysis, Mr. Hatem Yazidi and Dr. Moammar Ali AlTubi for providing me the technical knowledge in Matlab optimization, Mr.Mohammed Salim Adeeb in providing me the technical help in the laboratory, Ganesh Nagraj Patil, Dr.Lakkim Setty

Rao, and Dr.Santosh Madhav Walke for their help in the calculations of design mixing tank. Last but not the least, I would like to thank National University of Science and Technology-College of Engineering for giving me this opportunity to complete the project in their laboratories and supporting me by providing me with the required chemicals.

REFERENCES

1.Alaneme,K. et al .,2016.Corrosion Inhibitory Properties of Elephant Grass(Pennisetum Purpureum) Extract Effect on Mild Steel Corrosion in 1M HCl Solution :*Alexandria Engineering Journal*.[e-journal]. 55 (8). p. 1069-1076. Available from: https://www.sciencedirect.com/science/article/pii/S1110016816300242.[Accessed: 8th May 2019].

2.Ambrish,S. et al .,2012. The Effect of Environmentally Bening Fruit Extract of Shahjan(Moringa Olifera) on the Corrosion of Mild Steel in Hydrochloric Acid Solution:*Journal of Chemical Engineering Communications*.[e-journal]. 199 (6). p. 63-77. Available from:https://www.tandfonline.com/doi/abs/10.1080/00986445.2011.570390.[Accessed: 6th May 2019].

3.Blumberg,J.,et al.,2013.Carnberries and Their Bioactive Constituents in Human Health:*Journal of Advances in Nutrition*.[e-journal]. 12 (7). p. 618-632. Available from: https://academic.oup.com/advances/article/4/6/618/4595566.[Accessed: 1th May 2019].

4.Bobinaite,R.,Viskelis,P.&Venskutonis,P.,2015.Chemical Composition of Raspberry Cultivars:*International Journal of Molecular Sciences*.[e-journal]. 16 (8). p. 18642-18663. Available from: https://www.researchgate.net/publication/306158828_Chemical_Composition_of_Raspberry_Rubus_spp_Cultivars.[Accessed: 1th May 2019].

5.Cafferty,E.,2010.*Introduction to Corrosion Science*.United States:Springer.
Roberge,P.,2012.*Handbook of Corrosion Engineering*.2nd edition.United States:McGraw-Hill company.
6.Dynamixianc,2015.*Optimal Mixer Design* .[Online]. Available from: https://www.dynamixinc.com/optimal-mixer-design. [Accessed: 14th December 2018]

7.Ibryamoval,Y. et al .,2009.Design of Mixers for Chemical and Biochemical Industry: *Trakia Journal of Sciences*.[e-journal]. 8 (6). p. 6-11. Available from: http://tru.uni-sz.bg/tsj/Vol8.Suppl.3.2010/Y.Ibryamova.pdf.[Accessed: 13th December 2018].

8.Kaume, L. et al.,2012. The Blackberry Fruit A Review on Its Composition and Chemistry, Metabolism and Bioavailability and Health Benefits:*Journal of Agricultural and Food Chemistry*.[e-journal]. 37 (7). p. 5716-5727. Available from: https://www.ncbi.nlm.nih.gov/pubmed/22082199.[Accessed: 6th May 2019].

9.Kumar.A,2012. Inhibition of Mild Steel Corrosion in HCl Solution Using Pipali(Piper Longum)Fruit Extract:*Arabian Journal for Science and Engineering*.[e-journal]. 38 (9). p. 85-97. Available from: https://link.springer.com/article/10.1007/s13369-012-0409-9.[Accessed: 9th May 2019].

10. Michalska,A.& Lysiak.G .,2015.Bioactive Compounds of Bluberries:*International Journal of Molecular Sciences*.[e-journal]. 16 (8). p. 18642-18663. Available from: https://www.ncbi.nlm.nih.gov/pmc/articles/PMC4581264/.[Accessed: 1th May 2019].

11.*Mixing Tank*,2018.[Online]. Available from: https://www.conceptdraw.com/examples/agitator-tank-symbol.[Accessed: 19th November 2018].

12. Mycourses,2016. *Response Surface Modelling-RSM*.[Online]. Available from:https://mycourses.aalto.fi/pluginfile.php/364340/mod_resource/content/1/BOS_Lecture3_2016.[Accessed: 4th May 2019].

13. Ostovari, L. et al.,2009.Corrosion Inhibition of Mild Steel in 1M HCl Soluition by henna extract:*Journal of Corrosion Science*.[e-journal]. 32 (9). p. 1935-1949. Available from: https://www.sciencedirect.com/science/article/pii/S0010938X09002157. [Accessed: 6th May 2019].

14. Petropedia,2018. *Mixing Tank*.[Online]. Available from: https://www.petropedia.com/definition/7742/mixing-tank.[Accessed: 11th December 2018].

15.Raja,P. et al.,2008.Inhibitive Effect of Black Pepper Extract on the Sulphuric Acid Corrosion of Mild Steel:*Journal of Materials Letters*.[e-journal]. 62 (9). p. 1977-2979. Available from: https://www.sciencedirect.com/science/article/abs/pii/S0167577X08001316.[Accessed: 8th May 2019].

16. Sinnot,R.&Towler,G.,2009.*Chemical Engineering Design*.5th edition.India:Elsevier.

17. Xianghong,L. et al.,2009.Inhibition Effect of 6-benzylaminopurine on the Corrosion of Cold Rolled Ssteel in H_2SO_4 Soluition:*Journal of Corrosion Science*.[e-journal]. 51 (9). p. 620-634. Available from: https://www.sciencedirect.com/science/article/pii/S0010938X09000080. [Accessed: 6th May 2019].

590434 Verification of Firefighters' Heuristics and Predictive Model Development through Big Data Analysis

Abstract:

The heuristics accumulated in the field activities of firefighters are reviewed through big data analysis of fire occurrences and emergency responses in Gyeonggi-do Province, Korea and expanded as predictive models to be utilized for proper fire prevention activities according to time, day, and target through quantitative modeling. Empirical rules with high sympathy are collected through direct interviews with firefighters. Among these, three types of high-priority rules are compared and analyzed for big data such as fire occurrence and damage history in Gyeonggi-do in 2018. Big data comparison analyses are conducted, including the number of fires and damages that occurred in Gyeonggi-do in 2018. Furthermore, fire occurrence patterns by region, day of the week, time of day, and building type are derived. In addition to using empirical rules that have been validated through research, relatively inexperienced firefighters also can make better decisions by relying on refined quantitative predictive models replacing empirical rules, including local government and time-based factors that reflect big fire occurrence data.

CONVERTING SHELTER-IN-PLACE LOCATIONS TO ZERO VULNERABILITY SAFE HAVENS

J. A. Rau
MineARC Systems America, LLC
4850 W. Ledbetter Drive
Dallas, Texas 75236
USA

A. Young
MineARC Systems, Pty Ltd
274 Welshpool Rd
Welshpool, WA 6106
Australia

M. McDermott
High Performance Building Solutions
7415 Folger Dr.
Charlotte, NC 28226
USA

CONVERTING SHELTER-IN-PLACE LOCATIONS TO ZERO VULNERABILITY SAFE HAVENS

ABSTRACT

While it is not uncommon to find shelter-in-place or safe haven locations designated across chemical facilities, what varies is the level of protection offered at these sites. Functionality can range from merely a muster point (shelter-in-place) to a positively pressurized room (safe haven). Either way, when determining the room's acceptable performance criteria, the industry generally applies ERPG-3[1] or a vulnerability factor. An acceptable vulnerability factor is often deemed the probability of a fatality being less than one percent.

In 2017, MineARC Systems and High Performance Building Solutions (HPBS) began working with a leading energy company to convert five shelter-in-place spaces to safe havens, which included additional life support systems that would ensure zero percent vulnerability. The objective of this paper is to examine the process to harden the space against contaminated air intrusion and what critical life support systems are essential for converting a shelter-in-place location to a safe haven.

INTRODUCTION

Emergency shelter-in-place (SIP) forms an integral part of any plant's broader emergency response plan. Explosions, fires, and the release of smoke or other toxic gases are some of the types of incidents that can occur despite high levels of planning and safety precautions[2].

Numerous government agencies and other organizations recommend shelter-in-place as a measure to reduce harm in the event of a chemical release. Any building may be used as a temporary measure to reduce health risk from exposure to the toxic materials by simply closing windows and doors and turning off ventilation fans and air conditioning/heating systems.

Three factors govern the effectiveness of a SIP location: 1) the tightness of the building or leakage rate, 2) the concentration of the toxic gas, and 3) the release duration. The tightness of the building is primarily based on building construction; however, temperature differential and wind speed also play a factor. Identification of the maximum toxic concentration occurs during a Facility Siting Study (FSS). FSSs are a hazard analysis that defines Maximum Credible Event (MCE) scenarios using a consequence-based approach to produce a quantitative risk assessment. Determining potential consequences is a necessary step in the process of developing a comprehensive safety plan. The FSS determines the expected release duration based on stored inventories of toxic materials. The higher the toxic concentration and longer the release duration, the less effective sheltering in place becomes.

While occupants may feel safe sheltering-in-place from the external toxic hazard, typical buildings have high leakage rates. A typical wood framed constructed building can have 3-5 air changes per hour (ACH50[3]), rendering it ineffective against most chemical releases. Additionally, sealing to an acceptable airtightness can be time consuming and expensive and even with professional air sealing it is very difficult to air seal existing structures to an acceptable tightness.

A case study for the conversion of five shelter-in-place spaces to zero percent vulnerability safe havens is presented. There are considerations for effectively sealing existing infrastructure and ensuring positive pressure to ensure contaminants do not infiltrate safe haven locations along with the critical element of life support. The critical elements of life support that are of primary importance to human survival in a safe haven are:

- Positive pressure
- Breathable air supply
- Supplemental O_2
- Removal of accumulated CO_2

---

[1] Emergency Response Planning Guidelines (ERPGs) are values developed by the AIHA Guideline Foundation's Emergency Response Planning Committee. ERPG-3 is the maximum airborne concentration below which it is believed nearly all individuals could be exposed for up to 1 hour without experiencing or developing life-threatening health effects.

2 The CSB is an independent federal agency charged with investigating industrial chemical accidents. Over the last 10 years they have completed investigations into 15 separate chemical release incidents.

[3] ACH50 is the number of times the air volume in a building changes per hour at 50 Pascal of pressure. Commercial air leakage rates are commonly reported at 75 Pascals, but this leakage study was conducted at 50 Pascals based on the targets included in the FSS report.

- Cooling and dehumidifying

CASE STUDY

In 2017 MineARC Systems and its project partner HPBS began working alongside a major energy company to convert five shelter-in-place locations into zero percent vulnerability safe havens.

The process started with the energy company completing its FSS, which included plume modeling to define the chemical release concentrations at each building location along with ideal shelter-in-place locations[4] and internal minimum positive pressure[5]. This information was used by MineARC to calculate indoor concentrations and occupant vulnerability to highlight that shelter-in-place alone would result in the vulnerability of greater than zero percent.

From there, HPBS conducted air infiltration testing following ASTM E779-10 to determine an air leakage baseline (see Table 6) for rooms that were designated to be sealed (pre-weatherization). A site assessment was then completed to review each location and determine its suitability for sealing and conversion to a safe haven. Firstly, the measured baseline infiltration rates were compared to the FEMA 453 estimates based on construction type[6] (see Figure 2). All surfaces of the room were examined for ease of conventional air sealing and mechanical equipment sealing. Rooms that included false floors, large amounts of electrical conduit, cabinetry, computer servers or windows etc. were deemed difficult to seal (i.e. control rooms). Secondly, the building or rooms open floor space was checked to ensure it could accommodate the specified number of people in an emergency. FEMA 453 section 1.6 recommends $20ft^2$ per person for durations of less than 24 hours. With the FSS recommending 4 hours, it was deemed acceptable to have an unrestricted floor space of $6ft^2$ per person.

Manual weatherization (air sealing) of the rooms included installing a cleanroom suspended ceiling system with gaskets and tiles that met the air permeance rate for an air barrier material, cleanroom airtight light fixtures with gaskets, sealing around electrical and plumbing, air sealing all visible air pathways to make sure the interior gypsum board was functioning as the air barrier, installing low-leakage HVAC dampers, installing an emergency shut off for the mechanical system and adding door weather-stripping and sweeps. Post-weatherization air leakage tests were conducted to determine the air infiltration rate of the SIP spaces after completion of the manual air sealing, cleanroom suspended ceiling system and HVAC work. The weatherization air sealing methods improved the air barrier integrity for all the proposed SIP spaces by between 57% and 65% (refer to Table 6).

Aerosol-based envelope sealing was used to reduce leakage rates even further. AeroBarrier® is a building envelope sealing technology that simultaneously measures and seals buildings. AeroBarrier® works by forcing an atomized sealant into orifices under pressure gradually sealing leaks. Rooms were sealed during night shift with personnel able to re-access the rooms by morning. The AeroBarrier® air sealing methods improved the air leakage for all the SIP spaces between 7% and 88% over the post manual weatherization leakage rates (refer to Table 7). AeroBarrier® is not effective for sealing large holes so manual air sealing is required for certain locations prior to using AeroBarrier®.

MineARC engineers once again calculated infiltration concentration rates along with the vulnerability rate. As the probability of a fatality was still higher than zero, MineARC then specified its AirBANK Pressure System and necessary carbon dioxide and oxygen control equipment.

SAFE HAVEN LIFE SUPPORT EQUIPMENT

Primary Life Supporting Atmosphere

A safe haven needs to provide a reliable breathable atmosphere independent of external toxic gas concentrations. Human breathing consumes oxygen (O_2) and expels carbon dioxide (CO_2), carbon monoxide (CO), water vapor, and heat. At complete rest, an average person exhales air at 17% O_2 and 3.2% CO_2. This breathing rate results in a 'resting' O_2 uptake of 0.3L/min of O_2 and a resting CO_2 discharge of 0.24L/min (Brake, Fellows, & Bates, 1999). Taking into consideration the above values, a person in a closed system such as a shelter-in-place succumbs to CO_2 poisoning well before O_2

4 Shelter-in-place location is beyond the scope of this paper, but should be determined based on the chemical hazard, facility layout, and location of personnel.

5 Positive pressure requirements are based on average wind speed at the facility. As wind speed increases so does the SIP leakage rate.

6 The FEMA 453 estimates are intended to estimate filter sizing when ASTM E779 air leakage testing is not conducted.

asphyxiation. Venter et al. (1998) found an average oxygen consumption for 12 entrapped individuals of 0.44 liters per minute, at STP, over 24 hours (including sleeping). A figure of 0.5 liters per minute over a non-sleeping entrapment period of 12 hours is a prudent, conservative design value. This corresponds to a metabolic rate of about 160W or $80W/m^2$ for a typical person with a body skin area of about $2m^2$, which is equivalent to a breathing rate of about 12.5 liters per minute of fresh air.

Consequently, life support systems are needed within a safe haven that can ensure positive pressure keeps contaminants outside, along with replacing the O_2 consumed, and removing exhaled CO_2, moisture, and heat to maintain the atmosphere within acceptable threshold limits.

Positive Pressure

Positive pressure is integral to prevent the ingress of airborne contaminants that may compromise the breathable atmosphere for occupants. Establishing a Class 1 Safe Room (FEMA 453) requires the sufficient release of filtered air from an external source, to form a positively pressurized zone.

Buildings are typically constructed to provide natural or forced ventilation to allow permanent occupancy and potentially combined with large exposed windows and surfaces which can present challenges for ensuring protection during a chemical release. When determining adequate positive pressure within a space, several structural factors need to be considered, including:

- Shelter volume
- HVAC (Heating, Ventilating, and Air Conditioning)
- Door and window tightness
- Existing opening such as vents and elevator shafts
- Indoor and outdoor temperature
- Wind speed and building orientation
- The external screening of the structure

Building design and multiple ingress points suggest the conversion of a smaller interior room within a building can provide a higher level of protection, than the whole building, as it is less subject to wind and buoyancy forces which induce the infiltration of external hazards.

Breathable air from compressed air cylinders venting directly into the SIP forms a positive internal pressure. Determining the amount of air required to reach the necessary level of positive pressure involves measuring the areas internal volume and the relationship between the internal and external atmospheric conditions.

After reaching required ACH (air changes per hour) or tightness, a positive-pressure management system must be sized for the expected shelter duration.

Breathable Air Supply

If the shelter-in-place is airtight, a supply of breathable air is necessary to support occupants over the required shelter duration. Unventilated SIPs that are tightly sealed cannot be occupied for long periods without the risk of high carbon dioxide levels. A secondary source of breathable air must be circulated within the shelter to meet occupancy over the required amount of time.

Clean, breathable air is supplied to a chemical, biological and radiological (CBR) safe room in a few ways: by trapping air inside the room and minimizing the air exchange, passing contaminated air through a filter (not practical for high concentration chemical releases or for chemicals that cannot be filtered) to purify it as it is supplied to the room, and the supplemental supply of oxygen through high pressure compressed air cylinders.

The air supply must contain enough oxygen in its mix, and not contain contaminants like carbon monoxide, carbon dioxide, water, or any other toxins that poison safe haven occupants during entrapment.

Table 1 lists international standards for flowrates for breathable compressed air for self-contained breathing apparatus (SCBA), supplied air respirators (SAR) and refuge chambers.

Table 1 – Standards for acceptable air supply requirements

Air supply requirements	Relevant Standard
Open-circuit positive-pressure SCBA, including all components and accessories except the air cylinder (shell) must resist the permeation and penetration of HD and GB chemical agents when tested for a period of 6 hours on an upper-torso manikin connected to a breathing machine operating at an airflow rate of 40 liters per minute (lpm), 36 respirations per minute, 1.1 liters tidal volume.	NIOSH 42 CFR Part 84 Respiratory Protective Devices
The flow rate should be at least 85L/min (3cfm) per person to remove expired CO2 and CO but at least 255L/min (9cfm) to provide adequate cooling and maintain comfortable conditions. The air inflow is assumed to be at a temperature of 26.5°C and 65% relative humidity	ITA Report No. 14 – Guideline for the provision of refuge chambers in tunnels under construction, May 2014
For manually operated blower type air-hose respirators, the blower shall be capable of delivering air to the space enclosed by the facepiece at a flow rate of at least 85 L/min.	AS/NZS 17:16:2003 Respiratory Protective Devices

The NIOSH standard of 40L/min is for SCBA apparatus (i.e. mouthpiece or face mask) where dilution of CO_2 is not a concern. Conversely, the ITA and Australian standards specify 85L/min and are concerned with both oxygen consumption and carbon dioxide dilution. It is important to note that three times the amount of air is required to maintain comfortable conditions inside a refuge chamber, when compared to simply breathing (Health & Safety Executive UK, 2006).

Unless the positive pressure system is supplying breathable air at a rate that maintains oxygen and compressed gas levels within acceptable ranges (dependent on safe haven occupancy and duration), a CO_2 scrubber and supplementary oxygen supply are required.

Supplemental O_2

O_2 is an essential requirement for cellular activity, and therefore, the absence of O_2 for relatively short periods results in death. Humans within an enclosed environment consume O_2 and if not replaced, would result in an irrespirable atmosphere. An O_2 concentration of less than 11% poses a high risk of death.

Retaining the oxygen levels between acceptable limits is critical. Various international standards outline acceptable O_2 limits for a breathable environment; examples are shown in Table 2.

Table 2 - Standards for allowable O_2 levels

Minimum O_2 Content (% by volume)	Maximum O_2 Content (% by volume)	Relevant Standard
18.5%	23.0%	Mines Safety and Health Administration 30 CFR Parts 7 and 75 Refuge Alternatives for Underground Coal Mines; Final Rule
19.5%	23.5%	Occupational Safety and Health Administration 29 CFR 1910.146 Permit-Required Confined Spaces for General Industry
19.5%	23.5%	Australian Standard AS 2865–2009 Safe working in a confined space

OSHA and Australian Standards for working in a confined space specify an acceptable lower limit of 19.5%, which is consistent with many other standards around the world. The slightly lower MSHA limit of 18.5% is based on a person at rest, not working, therefore, not likely to have trouble in breathing; conditions that are likely to be present in a refuge chamber.

O_2 itself is not flammable but supports and accelerates combustion. OSHA defines O_2 enriched mixtures or atmospheres as those containing more than 23.5% O_2 by volume (OSHA, 2013). In O_2-enriched atmospheres, the reactivity of O_2 significantly increases the risk of ignition and fire.

Within a safe haven, pure O_2 must be utilized in conjunction with a CO_2 removal system.

Removal of Accumulated CO_2

Protection within a safe haven is substantially increased by filtering CO_2 from the air. CO_2 is being produced continuously by the cells of our body and is a byproduct of respiration. The inability to expel this gas or an environment with high concentrations of CO_2 leads to CO_2 poisoning. Breathing air with more than 3% CO_2 can induce dizziness, headaches, and vision disturbance. Prolonged exposure at greater than 6% could result in unconsciousness and death. Listed in Table 3 are examples of international standards of satisfactory CO_2 removal requirements for a breathable environment.

Table 3 – Acceptable CO_2 removal requirements

CO_2 Removal Requirements	Relevant Standard
CO_2 removal components shall remove CO_2 at a rate of 1.08 ft^3 (30.5 L)/h/person.	Mines Safety and Health Administration 30 CFR Parts 7 and 75 Refuge Alternatives for Underground Coal Mines; Final Rule
OSHA has established a Permissible Exposure Limit (PEL) for CO2 of 5,000 parts per million (ppm) (0.5% CO2 in air) averaged over an 8-hour workday (time-weighted average or TWA.)	Occupational Safety and Health Administration
5.2.8 Atmospheric monitoring. Atmospheric monitoring There should be the capability to monitor temperature continuously, and for carbon monoxide, carbon dioxide, and oxygen in the chamber during use in both externally supported or standalone modes. The concentration of carbon dioxide should not exceed 10000 ppm.	ITA Report No. 14 – Guideline for the provision of refuge chambers in tunnels under construction, May 2014

The CO_2 that is exhaled by the occupants within the safe haven must be removed by, constant flushing by an external breathable air source (e.g. compressed air cylinders used for positive pressure) or by the regeneration of the internal atmosphere with a CO_2 absorbent (scrubber).

Different regulations around the world, state varied human production rates between 30–50L per hour per person. When considering an acceptable production rate; it must be noted that the rate of breathing is primarily triggered by the CO_2 content of inspired air with higher CO_2 levels triggering faster respiration rates (Brake, Fellows, & Bates, 1999) (i.e. if the maximum allowable CO_2 concentration inside the safe haven is 1%, a lower production rate is acceptable compared to a 3% CO_2 limit).

If CO_2 removal mechanisms are inadequate, the occupants can develop CO_2 toxicity by re-breathing their own exhaled CO_2.

Cooling and Dehumidifying

Thermal tolerability with a safe haven should be addressed during the design phase. Assumptions must be made in preparation for safe haven design in regard to the number of personnel, activity level, available hydration and heat production. The minimum airflow to maintain a comfortable environment is set by taking these assumptions into account as well as the input temperature and humidity of supplied air.

The healthy human body maintains its internal temperature around 37°C. An essential requirement for continued normal body function is that the deep body temperature is maintained within the narrow 1°C limit (Epstein, Moran, 2006). A change of body temperature exceeding 1°C occurs only during illness or when environmental conditions surpass the body's ability to cope with extreme temperatures.

When a person is subject to extreme heat and humidity over some time, the person may experience a very rapid progression of symptoms or exhibit a few of the less significant symptoms before falling victim to more severe forms of heat-related illness.

The acceptable cooling requirements for underground refuge chambers are defined in various international standards, examples of which are listed in the below table.

Table 4 – Standards for acceptable cooling requirements

Cooling Requirements	Relevant Standard
Simulated emergencies, in which a full complement of people has occupied a refuge chamber for a significant period, indicate that humidity and temperature can increase very rapidly to potentially heatstroke-inducing levels. Refrigerated air conditioning is strongly recommended for both externally supported and stand-alone refuge chambers to counter this potentially dangerous problem. Inevitably, this places a heavy demand on the stand-alone power supply, but there are systems available that can cope, and at an acceptable cost.	Department of Mines, Industry Regulations and Safety, 2005 Refuge Chambers in Underground metalliferous mines – guideline: Safety and Health Division, Department of Industry and Resources Western Australia, 32 pp
Apparent temperature in the fully occupied refuge chamber must not exceed 95 °F.	Mines Safety and Health Administration 30 CFR Parts 7 and 75 Refuge Alternatives for Underground Coal Mines; Final Rule
It should be possible to maintain an internal temperature not exceeding 30 °C wet bulb globe temperature for at least 24 hours	ITA Report No. 14 – Guideline for the provision of refuge chambers in tunnels under construction, May 2014

MSHA regulations specify a maximum internal apparent temperature[7] of 95°F (35°C). NIOSH specifies an equivalent maximum wet bulb temperature of 28.9°C (NIOSH, 2007).

The temperature maintained inside the safe haven is determined by the metabolic heat liberated by the occupants, dissipated to the external environment by the heat transfer characteristics of the room or building. The external temperature primarily influences the heat transfer while the humidity is the combination of moisture produced by the occupants (via exhaled air and sweat) and water liberated by CO_2 scrubbing chemicals.

It is important to remember that in most cases, before sheltering-in-place, the air conditioning system will need to be shut down (unless it is a split system). For short durations, artificial cooling is not a critical requirement unless external temperatures are extraordinarily high, and/or the safe haven's volume is small compared to its occupancy.

INFILTRATION TESTING

Background

High Performance Building Solutions Inc. (HPBS) was contracted to perform an air infiltration study for spaces identified as candidates to be used as shelter-in-place (SIP) in the event of an accidental chemical release. This SIP strategy uses the existing structure and its indoor atmosphere to separate individuals from a hazardous outdoor atmosphere temporarily.

A series of pre-weatherization air leakage tests were conducted to determine the baseline air infiltration rate of potential SIP spaces. Manual air sealing, a cleanroom suspended ceiling system and low-leakage HVAC dampers were installed to reduce the overall air infiltration of the SIP spaces and to close off air leakage pathways from non-protected spaces in the building through the HVAC system. Post-weatherization air leakage tests were conducted to determine the air infiltration rate of the SIP space after the completion of weatherization and HVAC work. The SIP spaces were additionally air sealed using the proprietary AeroBarrier® system to further reduce infiltration rates.

Scope

Air sealing work was completed, and low-leakage HVAC dampers were installed to reduce the overall air infiltration of the SIP spaces and to isolate the HVAC from non-protected spaces in the building. Post-weatherization air leakage tests were conducted to determine the air infiltration rate of the SIP space after the completion of air sealing, suspended ceiling and HVAC work. AeroBarrier® air sealing was conducted on SIP spaces as a pilot project on September 18 through September 21, 2017. Post AeroBarrier® air leakage tests were conducted on September 21 and September 22, 2017, to determine the air infiltration rate of the SIP space after the completion of the AeroBarrier® sealing.

7 Apparent temperature is the temperature equivalent perceived by humans, caused by the combined effects of air temperature and relative humidity.

Following ASTM E 779-10, pressurization air leakage tests were conducted on the following spaces recommended in the Facility Siting Study (FSS).

Table 5: SIP specifications

SIP Space	Occupancy	Floor Area (Square Feet)	Volume (Cubic Feet)
SIP Space 1	40	990	10,481
SIP Space 2	6	955	8,210
SIP Space 3	5	391	3,917
SIP Space 4	10	1,697	16,968
SIP Space 5	40	293	2,637

Air Leakage Test Results

Positive pressure air leakage test results are reported in four different ways:

CFM4 is the quantitative volume of air passing through an enclosure under a pressure difference of 4 Pascals (equivalent to a 5mph wind). This is a measurable physical quantity and the value recommended in the FSS to eliminate toxic gas ingress into the SIP spaces.

CFM25 is the quantitative volume of air passing through an enclosure under a pressure difference of 25 Pascals (0.1-in w.g.). FEMA 453 requires Class 1 Safe Havens be pressurized to 25 Pascals. This is a measurable physical property.

CFM50 is the quantitative volume of air passing through an enclosure under a pressure difference of 50 Pascals (0.2-in w.g.).

ACH50 is the quantitative volume of air passing through an enclosure under a pressure difference of 50 Pascals (0.2-in w.g.) normalized by the enclosure volume. ACH is the abbreviation for air changes per hour. It is the number of times the air volume in a building changes per hour at 50 Pascals of pressure.

Table 6 below shows the CFM50 and ACH50 for each of the SIP spaces pre-weatherization and post-weatherization along with the percentage reduction in leakage. SIP Space 1 and SIP Space 5 did not have pre-weatherization results as initially the whole building was tested. Smaller spaces were selected as the whole building was too large to feasibly convert to a SIP space (re-testing of the new SIP spaces was not completed). The control room, SIP space 4, was abandoned in favor of a smaller adjacent room due to high leakage post manual weatherization and the computers that could not be shut down, which prevented the use of AeroBarrier® in the space.

Table 6: Post-weatherization leakage results

Safe Haven Space	CFM50		ACH50		% Reduction
	Pre-Weatherization	Post-Weatherization	Pre-Weatherization	Post-Weatherization	
SIP Space 1	N/A	1,871	N/A	10.7	N/A
SIP Space 2	3,623	1,560	26	11.4	57%
SIP Space 3	385	135	6	2.1	65%
SIP Space 4	N/A	N/A	N/A	N/A	N/A
SIP Space 5	N/A	655	N/A	19.4	N/A

The AeroBarrier® air sealing was prescribed a target. According to the AeroBarrier® simultaneous air leakage calculations, all the spaces met and surpassed this airtightness target; AeroBarrier® sealing was halted at this point. Table 7 shows the CFM50 and ACH50 for each of the SIP spaces as tested by HPBS post AeroBarrier® along with the percentage reduction. The HPBS results were utilized to calculate the amount of positive pressure required to convert the rooms to positive pressure safe havens.

Table 7: Post AeroBarrier® leakage results

Safe Haven Space	CFM50		ACH50		% Reduction
	Post-Weatherization	Post AeroBarrier®	Post-Weatherization	Post AeroBarrier®	
SIP Space 1	1,871	455	10.7	2.6	76%
SIP Space 2	1,560	182	11.4	1.3	88%
SIP Space 3	135	125	2.1	1.9	7%
SIP Space 4	N/A	173	N/A	3.3	N/A
SIP Space 5	655	138	19.4	3.1	79%

The AeroBarrier® air sealing methods improved the air barrier integrity for all the proposed SIP spaces between 7% and 88% over the post manual weatherization leakage rates. Air leakage was still observed behind many electrical penetrations including HVAC thermostats, and fire and smoke alarms. These were covered with plastic to protect them from the AeroBarrier® Sealant but had hidden holes behind them.

Figure 1 displays the post AeroBarrier® leakage rate in comparison to the FEMA 453 estimates for air leakage rates based on construction types. FEMA 453 estimates are reported in square feet of floor area at 25 Pascals and do not take into consideration the ceiling height and volume of the space.

Figure 1: Post AeroBarrier® results compared to FEMA 453 estimates

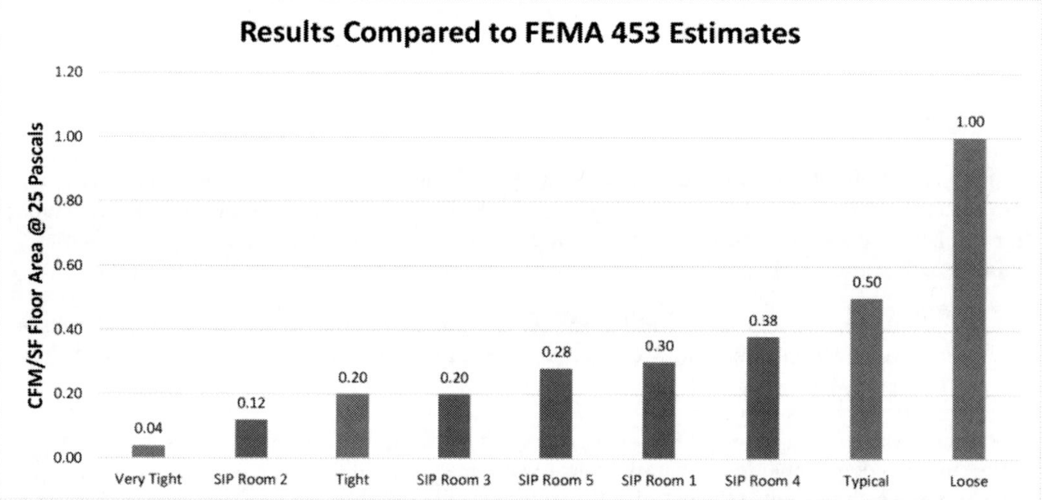

Figure 2 is an estimated leakage rate based on construction type to estimate the filtration and positive pressure system size under the FEMA 453 standard. All the SIP locations were between tight and typical except for SIP Room 2, which was a small prefabricated metal constructed portable building.

Figure 2: Estimated leakage rate per square foot of floor space to achieve 0.1 iwg (25Pa) overpressure

Construction Type	cfm per square foot of floor area
Very tight: 26-inch thick concrete walls and roof with no windows	0.04
Tight: 12-inch thick concrete or block walls and roof with tight windows and multiple, sealed penetrations	0.20
Typical: 12-inch thick concrete or block walls with gypsum wall board ceilings or composition roof and multiple, sealed penetrations	0.50
Loose: Wood-frame construction without special sealing measures	1.00

Table 8 shows the combined AeroBarrier®, and manual air sealing methods improved the airtightness for all the proposed SIP spaces between 68% and 95% over the original leakage rates.

Table 8: Air infiltration improvement for SIP spaces

Safe Haven Space	ACH50			% Reduction
	Pre-Weatherization	Post-Weatherization	Post AeroBarrier®	
SIP Space 1	N/A	10.7	2.6	76%
SIP Space 2	26.5	11.4	1.3	95%
SIP Space 3	5.9	2.1	1.9	68%
SIP Space 4	N/A	N/A	3.3	N/A
SIP Space 5	N/A	19.4	3.1	84%

Figure 3 compares the ACH50 pre-weatherization, post-weatherization and AeroBarrier® air infiltration results.

Figure 3: Pre-weatherization, post-weatherization and AeroBarrier® comparison

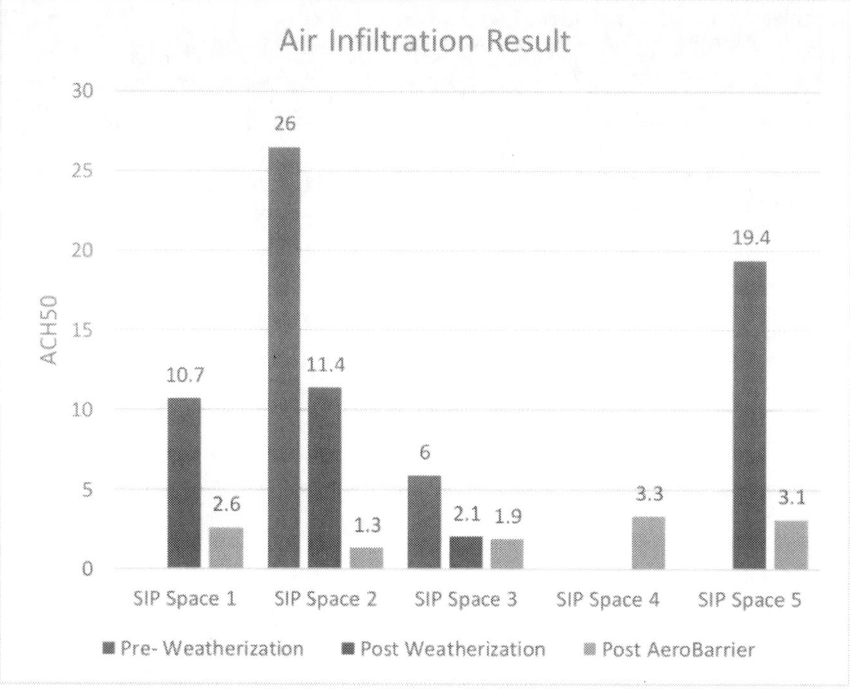

Life Support Systems

MineARC re-calculated the vulnerability for each SIP space based on the post AeroBarrier® leakage rates. All SIP spaces still had a vulnerability of 100% and therefore required the installation of independent life support systems to be deemed zero percent vulnerability safe havens.

Figure 4 shows the toxic gas concentration, duration of the release, and vulnerability for each SIP space. As the largest MCE was a full tank rupture the toxic concentrations were extremely high, but the duration was short at one minute.

Figure 4: SIP space vulnerability post AeroBarrier®

	TOXIC GAS				
Safe Haven Location	Worst Case Threat Scenario	Toxic External [ppm]	Duration of Reslease [min]	ACM	Vulnerability
SIP Space 1	Tank Rupture	585,000	1	0.032	100%
SIP Space 2	Tank Rupture	652,000	1	0.042	100%
SIP Space 3	Tank Rupture	660,000	1	0.007	100%
SIP Space 4	Tank Rupture	412,000	1	0.042	100%
SIP Space 5	Tank Rupture	600,000	1	0.028	100%

The leakage rate at 4 Pascals of pressure difference is slightly more than would occur under a 5 mile per hour wind used in the dispersion model in the FSS report based on climatic data for the facility (to maintain positive pressure and neutralize air infiltration). Post-weatherization infiltration rates in cubic feet per minute at 4 Pascals of pressure for each of the SIP spaces is shown in Table 9.

Table 9: Post AeroBarrier® air leakage results at CFM4

Safe Haven Space	Volume (Cubic Feet)	CFM4
SIP Space 1	10,481	99
SIP Space 2	8,210	32
SIP Space 3	3,917	24
SIP Space 4	3,110	45
SIP Space 5	2,637	21

The air leakage test results provide a mathematical formula that can be used to calculate the amount of breathable compressed air cylinders ($437ft^3$) that must be added to each candidate SIP space to maintain a 4 Pascal positive pressure to overcome a 5mph design wind coincident with a toxic release.

MineARC Systems was contracted to install life support systems to convert the SIP spaces to positive pressure safe havens. The MineARC AirBANK Positive Pressure System provides rapid pressurization, which is activated and maintained using the AirBANK Control via a simple human-machine interface (HMI) touch screen. MineARC's integrated Aura-FX Gas Monitor ensures breathable air automatically remains within acceptable limits. Additionally, the installation of supplementary oxygen and carbon dioxide scrubbing systems in safe havens, are as needed.

Figure 5 shows the number of 437 cubic feet cylinders required for each safe haven location. The minimum breathable air per person is 2.5cfm (70L/min) per person based on MineARC's experience and testing within the refuge chamber industry.

Figure 5: Quantity of cylinders required for 4Pa positive pressure

	SAFE HAVEN SPECIFICATIONS		POSITIVE PRESSURE (@ 4Pa)						
Safe Haven Location	Occupancy	Duration	CFM	CFM per Floor Space	ACH	No. Cylinders	No. Cylinder Racks	Breathable Air per Person [CFM]	Breathable Air per Person [LPM]
Safe Haven Space 1	40	4	99	0.10	0.6	55	5	2.5	70.8
Safe Haven Space 2	6	4	32	0.03	0.2	18	2	5.4	152.9
Safe Haven Space 3	5	4	24	0.06	0.4	14	2	4.8	135.9
Safe Haven Space 4	10	4	45	0.14	0.9	25	3	4.5	127.4
Safe Haven Space 5	40	4	21	0.07	0.5	12	2	0.6	17.0

The maximum CO_2 concentration is at 1% and O_2 between 19.5% and 23.5% based on OSHA limits (refer to Table 2). Due to the volume and number of people in Safe Haven Space 5, it required supplemental oxygen and carbon dioxide scrubbing to maintain the internal gas concentrations within specified limits. Three sodium chlorate oxygen candles and two AirGEN CO_2 scrubbers were included in the design.

Figure 6: Internal gas concentrations in safe havens

	SAFE HAVEN SPECIFICATIONS		POSITIVE PRESSURE (@ 4Pa)			O2 & CO2 CONCENTRATIONS (@ 4Pa)			
Safe Haven Location	Occupancy	Duration	CFM	CFM per Floor Space	ACH	Indoor O_2 %	No. O_2 Candles	Indoor CO_2 %	No. AirGEN Scrubbers
Safe Haven Space 1	40	4	99	0.10	0.6	20.45	0	0.50	0
Safe Haven Space 2	6	4	32	0.03	0.2	20.84	0	0.17	0
Safe Haven Space 3	5	4	24	0.06	0.4	20.77	0	0.23	0
Safe Haven Space 4	10	4	45	0.14	0.9	20.66	0	0.31	0
Safe Haven Space 5	40	4	21	0.07	0.5	18.67	3	2.04	2

In 2018 MineARC completed installation and training on the five safe haven spaces. Each safe haven was tested during the commissioning process to ensure it could maintain acceptable positive pressure.

CONCLUSIONS

The conversion of shelter-in-place (SIP) locations to zero percent vulnerability safe havens is a four-step process consisting of the following:

1) Identifying the correct SIP location based on ease of sealing and small volume based on the required occupancy.
2) Manual air sealing of SIP spaces.
3) AeroBarrier® sealing of SIP spaces.
4) Installation of life support equipment.

Identifying SIP spaces that can seal from the outset is crucial to a successful result. In the case study, three of the five original locations were abandoned in favor of rooms in adjacent locations due to high leakage rates or reducing the size to make the project more cost effective. Control rooms pose the most difficulty in trying to seal due to the false floor and a myriad of cable tray and pass-throughs installed for the computer systems. The smaller the room and the more robust the construction (i.e. prefabricated metal or concrete buildings), the better it seals.

Manual air sealing needs to be overseen by someone with familiarity in building tightness. While most of the tasks are general construction, standard contractors are not trained to work at the level of precision needed to ensure proper sealing. Even a tiny orifice or ceiling tile not sealed correctly can ruin a leakage test result. HPBS and MineARC Systems have developed a system for ensuring airtightness in construction methodology and by avoiding un-sealable spaces or spaces that are not cost effective to seal.

For future construction projects, incorporation of SIP spaces should occur during the design phase. Air barrier commissioning is standard in new construction and should include the following tasks at a minimum: design review of the plans with specifications for the air barrier and mechanical system - specifically as these relate to spaces designated for use as safe havens; quality assurance inspections during construction to ensure a continuous air barrier is in place; and final air leakage testing of the continuous air barrier to ensure the leakage rate of the safe haven spaces are within acceptable limits.

Complementary life support equipment is essential to sustain life within an airtight environment. For personnel to remain within the sealed safe haven, the environment must be habitable. Air is to be free of toxins such as carbon dioxide and acceptable levels of oxygen maintained. Positive pressure systems reduce the ingress of toxins and provide some or all of the required source of breathable air. Additional equipment such as compressed air, oxygen cylinders and air scrubbers are used to meet the shortfall. Furthermore, internal conditions must consider the human condition, supporting workable body temperature and humidity.

Reducing the risk of harm to personnel is a priority for emergency response planning. While the likelihood of an MCE occurring may be low, the risk to on-site personnel in the event of an incident is high. Converting shelter-in-place (SIP) locations to zero vulnerability safe havens improves emergency response.

REFERENCES

ASTM E779-10, *Standard Test Method for Determining Air Leakage Rate by Fan Pressurization*, ASTM International, West Conshohocken, PA, 2019, www.astm.org

Brake, Fellow, & Bates (1999). *Criteria for the design of emergency refuge stations for an underground metal mine*. Aus. IMM Proceedings, No.2, 1999, pp 1–7.

Department of Mines and Petroleum, 2013, *Refuge chambers in underground mines — guideline: Resources Safety*, Department of Mines and Petroleum, Western Australia, 41 pp.

Epstein, Y., & Moran D. (2006). *Thermal Comfort and the Heat Stress Indices,* Industrial Health, 44, 388–398.

FEMA 453. (2006). *Design Guidance for Shelters and Safe Rooms.* Federal Emergency Management Agency

Health & Safety Executive UK (2008). *Guidance and Information on the role and design of safe havens in arrangements for escape from mines.* Retrieved from http://www.hse.gov.uk/pubns/mines08.pdf

ITA Working Group 5 (2016). *Provisions for refuge chambers in tunnel projects under construction.* ITA Report N°014 - V2 / March 2018

Mines Safety and Health Administration (2008). *30 CFR Parts 7 and 75 Refuge Alternatives for Underground Coal Mines; Final Rule.*

National Institute for Occupational Safety and Health (2007). *Research Report on Refuge Alternatives for Underground Coal Mines.* Retrieved from http://www.cdc.gov/niosh/docket/archive/pdfs/NIOSH-125/125 ResearchReportonRefugeAlternatives.pdf

Occupational Safety & Health Administration (2013). *Permit-required confined spaces.* Retrieved from http://www.osha.gov/pls/oshaweb/owadisp.show_document?p_table=STANDARDS&p_id=9797

Suming Du. (2014). *On the effectiveness of shelter-in-place as a measure to reduce harm from atmospheric releases. Journal of Emergency Management, Vol. 12, No, 3, May/June 2014.*

Venter. J. (1998). *Portable Refuge Chambers: Air or Tomb in Underground Escape Strategies.* School for Mechanical and Materials Engineering, Potchefstoom University, Republic of South Africa

Worksafe Australia National Standard (2005). *Australian Standard AS: NZS 1716:2003 Respiratory protective devices.*

Yantosik. G. (2006). *Shelter-in-Place Protective Action Guide*. The Chemical Stockpile Emergency Preparedness Program (CSEPP). U.S. Department of Energy

589008 How the Field of Human Factors Helps Solve the Riddle of Serious Injuries and Fatalities

Abstract:

Process and occupational safety are often treated as distinct endeavors, with the former tending to focus on facility system function and the latter on individual worker wellbeing. Though other differences certainly exist between the two, both endeavors are generally purposed to prevent incidents that can cost lives, health, time, and money, in addition to employee morale and public trust. The overarching thesis of this paper is that the field of Human Factors is uniquely positioned to effect positive change in both of these major safety categories.

The backdrop for our argument is the recent finding that while injury and illness incidence rates published by the Bureau of Labor statistics have decreased in recent years, the occurrences of serious injuries and fatalities have declined at a significantly slower rate. We discuss possible explanations for this finding but note that regardless of the preferred explanation, the trend should motivate scientists, engineers, and other safety professionals to revisit long-held accident causation theories holding that near misses – these are incidents that could have resulted in serious injuries and fatalities but did not – are precursors of future serious incidents. While this idea still has merit, the just-mentioned statistics warrant reconsideration of the relations between serious and less-serious incidents to explain why the rates of the two are not decreasing in lockstep. We suggest these data indicate additional (human) factors are involved.

To address this riddle and to support our general thesis, we (a) review core tenets of process safety, occupational safety, and the field of Human Factors, (b) point out that the human is always an element common to both process and occupational safety contexts, and (c) discuss how preventing or mitigating human error can benefit both. We offer specific examples from both safety contexts to show how Human Factors perspectives and techniques can help. Examples touch on standards, policies, and other administrative controls (e.g., procedures), human-machine interfaces of both workplace hardware (e.g., valves) and software (e.g., SCADA systems), environments (e.g., physical, regulatory, and cultural), and communications between personnel (e.g., elements of team resource management). We conclude that fundamental Human Factors considerations can help prevent and mitigate human performance gaps and therefore increase the rate of reduction in serious injuries and fatalities in both process and occupational safety.

589233 Initial Conditions Used in Atmospheric Dispersion Models:Issues with Commonly Used Assumptions for Gas Jets and Plumes

Abstract:

When simulating jets and plumes, all atmospheric dispersion models require initial conditions that appropriately parameterize the behavior of a gas release to the atmosphere. Often, process conditions are such that loss of primary containment of a gas will result in choked flow conditions. A great deal of experimental and theoretical consideration has been given to the flow from stagnation conditions to choked flow conditions, but less experimental work has centered on describing the flow through the depressurization process to atmospheric pressure where atmospheric dispersion models would be applicable. Traditional modeling of the depressurization process has taken the simplified approach of assuming a (depressurized) "top hat" profile (laterally constant) with application of the principles of conservation of mass, momentum, and energy. Such assumptions are known to result in predicted depressurized velocities that exceed the (sonic) velocity at the point of loss of containment, and because of the energy balance, the associated temperatures are significantly less than temperatures at the choked conditions. Physically, these effects could possibly self-cancel because as jet velocity slows, the decrease in kinetic energy of the jet will increase the jet temperature. However, no atmospheric dispersion model used for routine hazard assessment purposes accounts for kinetic energy in the energy balance. The traditional modeling approaches of the depressurization process are reconsidered based on analysis of available data, and an alternative approach to modeling the depressurization process is proposed. Example cases showing the important differences between the two approaches are discussed.

Fault Tree Uncertainty Analysis

Raymond "Randy" Freeman
S&PP Consulting
12303 Lake Shore Ridge
Houston, TX 77041
Cell: 713 408 0357
Rafree@yahoo.com

For Presentation at the 2020 Spring National AIChE Meeting
Houston, Texas

March 30, 2020

ABSTRACT

Fault tree analysis (FTA) is a widely used methodology in the process industries. FTA is used for the development of failure mechanisms, computation of failure frequencies and the determination of the probability of failure on demand of safety systems. Much of the data used in a FTA study are uncertain. For example, the failure rate of a pump is often not known with great precision. Likewise the failure rates of instrumentation are often known only within some defined limits. The common practice, used by analysts in the quantification of a fault tree, is to use the most likely or best guess as to the needed failure rate data. The use of best guess values as data inputs to the quantification of a fault tree creates uncertainty in the computed results.

This paper presents a general methodology for the determination of the impact of uncertainty on the results of a fault tree study. The general methodology is based on the mathematics of propagation of error and variance contribution analysis. For a fault tree that contains 10 event minimal cut sets, the methodology will compute the variance in the probability (or frequency) of the top event with a maximum error of 5%. For 5 event minimal cut sets the error in variance of the top event probability (or frequency) will be no more than 2%. An example is presented to illustrate the application of the fault tree uncertainty analysis methodology to a real world problem.

Keywords

Quantitative Risk Assessment, Fault Tree Analysis, Uncertainty Analysis

1. Background

Fault tree analysis is a widely used method for representation of failure mechanisms in chemical plants. The methodology is described in the CCPS CPQRA guideline book [Ref. 1] and is extensively discussed in Chapter 9 of Lees [Ref. 2]. Lacking in these discussions is a systematic method to quantify the uncertainty of the results of the fault tree calculations.

Fault tree analysis has also been used in the aircraft and aerospace industries, nuclear power plant review and is recognized by the OSHA Process Safety Management (PSM) standard [Ref. 3] as a method for process hazards analysis.

2. Review of Error Propagation and Variance Contribution Analysis (VCA) Methodology

The mean and variance of a function of random variables can be approximated using the method described by Haugen [Ref. 4] and applied by Freeman [Ref. 5, 6, 7]. Define an arbitrary function of a set of random variables, x_i, as:

Let

$$Y = F(x_i) \qquad \text{(Eq 1)}$$

The mean of Y can be estimated using the following approximation:

$$E(Y) = F[\,E(x_i)] \qquad \text{(Eq 2)}$$

Where:

E(Y) = expected value of random variable Y = mean of Y

$E(x_i)$ = expected value of random variable x_i = mean of x_i

The variance of Y can likewise be estimated as:

$$V(Y) = \sum_{i=1}^{n} \left[\frac{\partial Y}{\partial x_i}\right]^2 V(x_i) \qquad \text{(Eq 3)}$$

Where:

$V(Y)$ = variance of random variable Y as defined above in Equation 1

$V(x_i)$ = variance of random variable x_i as defined above in Equation 1

Note that the variance is simply the square of the standard deviation. Using the variance will simplify the mathematics that is described below. The contribution of each independent random variable to the overall variance in the function is:

$$V(Y \text{ from } x_i) = \left[\frac{\partial Y}{\partial x_i}\right]^2 V(x_i) \qquad \text{(Eq 4)}$$

The relative contribution of each term to the overall variance V(Y) is a measure of the importance in the uncertainty in the particular random variable, x_i. In effect, this is a sensitivity analysis combined with a uncertainty evaluation. The variance contribution combines the sensitivity in the answer to changes in the uncertain random variable, x_i, with a measure of the uncertainty in the random variable, x_i. The overall variance in Y is found by summing the sensitivity weighted variances from each random variable.

Numerical Estimate of the Sensitivity

Let us return to the fundamental definition of the derivative.

As before, set y = F(xi)

$$\frac{\partial y}{\partial x} = \lim_{\Delta xi \to 0}\left[\frac{F(xi+\Delta xi) - F(xi)}{\Delta xi}\right] \qquad \text{(Eq 5)}$$

We are interested in a numerical estimate of the derivative. Therefore, we will make the following approximation.

$$\frac{\partial y}{\partial x} \sim \left[\frac{F(xi+\Delta xi) - F(xi)}{\Delta xi}\right] \qquad \text{(Eq 6)}$$

Linear Functions

Let us examine the approximation for a simple linear function.

Let Y = AX + B (Eq 7)

What is the sensitivity of Y to a change in variable X? By simple calculus:

$$\frac{\partial y}{\partial x} = A \qquad \text{(Eq 8)}$$

We can also find the sensitivity using the numerical approximation as:

$$\frac{\partial y}{\partial x} \sim \left[\frac{F(xi+ \Delta xi)- F(xi)}{\Delta xi}\right] \qquad \text{(Eq 9)}$$

$$\frac{\partial y}{\partial x} \sim \left[\frac{(A(xi+ \Delta xi)+ B)-(A(xi)+ B)}{\Delta xi}\right] \qquad \text{(Eq 10)}$$

$$\frac{\partial y}{\partial x} \sim \left[\frac{(A(xi+ \Delta xi))-(A(xi))}{\Delta xi}\right] \qquad \text{(Eq 11)}$$

$$\frac{\partial y}{\partial x} \sim \left[\frac{((A\, xi+A\, \Delta xi))-(A(xi))}{\Delta xi}\right] \qquad \text{(Eq 12)}$$

$$\frac{\partial y}{\partial x} \sim \left[\frac{A\, \Delta xi}{\Delta xi}\right] \qquad \text{(Eq 13)}$$

$$\frac{\partial y}{\partial x} = A \qquad \text{(Eq 14)}$$

Note that for a simple linear equation the numerical result for the sensitivity is exactly the same as the analytical expression. There is no restriction on the size of the perturbation (Δxi) used in the calculations. Previously, Freeman [Ref 7] has suggested using a 10% perturbation in xi to do the sensitivity calculations. However any size of perturbation will work and generate the correct sensitivity.

This is a general result. As long as the function can be expressed as a linear function of a number of variables, xi, the sensitivity can be directly calculated by a simple perturbation of the function. The resulting sensitivity will be exactly the same as would be obtained using analytical methods.

Non-Linear Expressions

In the quantification of a fault tree, the top event frequency is computed using failure rates of the component devices and time that measures the interval between system validations. In a minimal cut set of two or more basic events, the time variable may be the same. For example the test interval for a valve may be the same for a pressure transmitter. A minimal cut set involving both the valve and the pressure transmitter would create a 2rd order term in the test interval. If the test interval is uncertain, how do we compute the sensitivity of the top event frequency with respect to the test interval? We will now explore how to deal with this type of problem when completing the uncertainty analysis.

As before, let

$$y = F(xi) = Axi^n + B \qquad \text{(Eq 15)}$$

What is the sensitivity of Y to a change in variable X? By simple calculus:

$$\frac{\partial y}{\partial x} = nAxi^{n-1} \qquad \text{(Eq 16)}$$

Let us evaluate the numerical approximation for this case. If n=2, the function of interest is:

$y = F(xi) = Axi^2 + B$ (Eq 17)

From basic calculus, the exact analytical expression for the sensitivity is:

$$\frac{\partial y}{\partial x} = 2\,A(xi) \qquad \text{(Eq 18)}$$

If xi = 1

$$\frac{\partial y}{\partial x} = 2\,A \qquad \text{(Eq 19)}$$

Again the approximation of the derivative is:

$$\frac{\partial y}{\partial x} \sim \left[\frac{(A(xi+\Delta xi)^2+B)-(Axi+B)}{\Delta xi}\right] \qquad \text{(Eq 20)}$$

$$\frac{\partial y}{\partial x} \sim \left[\frac{F(xi+\Delta xi)-F(xi)}{\Delta xi}\right] \qquad \text{(Eq 21)}$$

Inserting the equation for F(xi) we obtain:

$$\frac{\partial y}{\partial x} \sim \left[\frac{(A(xi+\Delta xi)^2+B)-(Axi+B)}{\Delta xi}\right] \qquad \text{(Eq 22)}$$

$$\frac{\partial y}{\partial x} \sim \left[\frac{(A\,(xi^2+2\Delta xi+\Delta xi^2)\,+B)-(Axi+B)}{\Delta xi}\right] \qquad \text{(Eq 23)}$$

$$\frac{\partial y}{\partial x} \sim \left[\frac{A\,(xi^2+2\Delta xi+\Delta xi^2)\,-Axi}{\Delta xi}\right] \qquad \text{(Eq 24)}$$

$$\frac{\partial y}{\partial x} \sim A\left[\frac{(xi^2+2\Delta xi+\Delta xi^2)\,-xi}{\Delta xi}\right] \qquad \text{(Eq 25)}$$

Setting xi =1, and using a 10% perturbation as $\Delta xi = 0.1\,xi = 0.1$

$$\frac{\partial y}{\partial x} \sim A\left[\frac{(1^2+(2)(0.1)1+0.1^2\,)\,-1}{0.1\,(1)}\right] \qquad \text{(Eq 26)}$$

$$\frac{\partial y}{\partial x} \sim A\left[\frac{(2)(0.1)+\,0.01}{0.1}\right] \qquad \text{(Eq 27)}$$

$$\frac{\partial y}{\partial x} \sim A\left[\frac{0.21}{0.1}\right] \qquad \text{(Eq 28)}$$

$$\frac{\partial y}{\partial x} = 2.1\,A \qquad \text{(Eq 29)}$$

Note that the numerical estimate (Eq 29) is approximately 5% greater than the analytical value (Eq 19). As shown in Table 1, for higher order equations the error will increase as the size of the perturbation becomes more important. Table 2 shows the impact of reducing the perturbation size to 1%. For 10th order variables (say a minimal cut set with 10 basic events) the error is 5%. For almost all engineering evaluations, a maximum error of 5% should be adequate.

For cases where the test frequency, repair time, or mission time appears in a minimal cut set 3 or more times, I suggest that the following perturbation in time be used to compute the sensitivity of the top event:

$$\Delta time = 0.01\ time \quad \text{(Eq 30)}$$

The sensitivity of the top event frequency to all uncertain variables can be computed using a perturbation of 1%.

$$\Delta xi = 0.01\ xi \quad \text{(Eq 31)}$$

3. Fault Tree Analysis

Fault trees are often used to analyze potential failures of an engineered system. A fault tree is a failure logic diagram that shows the relationship of device or system failures leading to an accident. Fault tree analysis is covered in detail in the CCPS CPQRA book [Ref. 1]. A fault tree is constructed by starting with a top event or condition and asking the question:

What conditions lead to this outcome?

Each condition in the next layer down is then analyzed by asking the same question. What conditions lead to this outcome? The logic of how these intermediate conditions (termed intermediate events) are related is expressed using AND and OR logic gates. An AND gate output is true if and only if (iff) all of the inputs to the AND gate are all true simultaneously. An OR gate output is true iff one of the inputs to the OR gate is true.

Once the fault tree logic structure is created, numerical data can be used to calculate the frequency or probability of the top event of concern occurring. For fault trees where a basic event is found in multiple branches of the tree, such as power failure, the logic structure of the tree must be simplified to remove the impact of the repeated events on the top event of concern. This simplification is done using Boolean Algebra. Each resulting grouping of basic events can cause the top event to occur. These groupings of basic events are called minimal cut sets. As a simple example consider the Boolean Algebra representation of a fault tree :

$$T = BE11 + BE1 \cdot BE2 + BE1 \cdot BE3 + BE2 \cdot BE9 + BE3 \cdot BE7 \quad \text{(Eq 32)}$$

Where:

T = Fault Tree Top Event

BE11 = Basic Event 11

BE1 = Basic Event 1

BE2 = Basic Event 2

BE3 = Basic Event 3

BE7 = Basic Event 7

BE9 = Basic Event 9

+ = Boolean Algebra Addition Symbol

• = Boolean Algebra Multiplication Symbol

For this example the top event T, the minimal cut sets are:

BE11

BE1 BE2

BE1 BE3

BE2 BE9

BE3 BE7

For example the existence of basic events BE1 and BE3 guarantees that the top event T will occur. Publically available standard computer software such as SAPHIRE [Ref. 8] may be used to find the minimal cut sets of a large fault tree.

The representation of the top event in equation 32 can now be used to compute the probability or frequency of occurrence of the top event. If the devices describe by the basic events are non-repairable and are tested after a period of time TI, the probability of top event, T, may be calculated as:

$$\text{Prob (T)} = \lambda_{11}\ TI/2 + (\lambda_1\ TI/2)(\lambda_2\ TI/2) + (\lambda_1\ TI/2)(\lambda_3\ TI/2) + (\lambda_2\ TI/2)(\lambda_9\ TI/2) + (\lambda_3\ TI/2)(\lambda_7\ TI/2) \qquad \text{(Eq 33)}$$

Where:

λ_{11} = failure rate of device 11 in basic event BE11, failures per unit time

λ_1 = failure rate of device 1 in basic event BE1, failures per unit time

λ_2 = failure rate of device 2 in basic event BE5, failures per unit time

λ_3 = failure rate of device 3 in basic event BE3, failures per unit time

λ_7 = failure rate of device 7 in basic event BE7, failures per unit time

λ_9 = failure rate of device 9 in basic event BE9, failures per unit time

TI = test interval, time

Equation 33 assumes that the device failure rates are small and the product λ TI < 0.1. The other major assumption is that the values of all of the failure rates and test intervals are known. In reality the failure rate data are not known exactly with only a range or probability distribution representing the state of knowledge.

4. Example Problem

The following example is taken from the ISA technical report on fault tree analysis [Ref. 9]. This example should not be considered as a recommendation by either ISA or AIChE. The example is presented to allow for the demonstration of the methods that can be used in a fault tree study to quantify the uncertainty in the resulting calculated failure frequency or failure probability.

Consider the interlock on the intermediate storage tank T101 shown on Piping and Instrument Diagram in Figure 1. The interlock block diagram is shown in Figure 2. The interlock is intended to prevent an abnormal condition in the tank which could lead to a uncontrolled release of the tank contents. The process hazards analysis team has recommended that the interlock be designed to safety integrity level 2 (SIL 2) with a target probability of failure on demand (PFD) of no greater than 1E-2 or a risk reduction factor (RRF) of 100. Does the proposed interlock shown on Figure 1 meet the SIL 2 target? What is the uncertainty in the predicted PFD of the interlock?

A fault tree for the failure of this interlock is shown in Figure 3. A minimal cut set analysis has been completed and the resulting minimal cut sets are presented in Table 3. We can now write the Boolean Algebra equation that represents this fault tree as:

$$T = PE + TS1 \cdot TS2 + LS1 \cdot LS2 + FT1 \cdot FT2 + FT2 \cdot FT3 + FT1 \cdot FT3 + BV1 \cdot BV2 + BV1 \cdot SOL1 + BV2 \cdot SOL2 + SOL1 \cdot SOL2 + PT1 \cdot PT2 \qquad \text{(Eq 34)}$$

Assuming that the interlock is non-repairable until tested at time TI and using lamda-time (λT) approximation of the failure rate, the Boolean representation of the fault tree may now be converted to a failure probability model as:

$$\text{Prob}(T) = \text{Prob}(PE) + (\lambda_{TS1}\, TI/2)(\lambda_{TS2}\, TI/2) + (\lambda_{LS1}\, TI/2)(\lambda_{LS2}\, TI/2) + (\lambda_{FT1}\, TI/2)(\lambda_{FT2}\, TI/2) + (\lambda_{FT2}\, TI/2)(\lambda_{FT3} TI/2) + (\lambda_{FT1}\, TI/2)(\lambda_{FT3}\, TI/2) + (\lambda_{BV1}\, TI/2)(\lambda_{BV2}\, TI/2) + (\lambda_{BV1}\, TI/2)(\lambda_{SOL1} TI/2) + (\lambda_{BV2}\, TI/2)(\lambda_{SOL2}\, TI/2) + (\lambda_{SOL1}\, TI/2)(\lambda_{SOL2}\, TI/2) + (\lambda_{PT1}\, TI/2)(\lambda_{PT2}\, TI/2) \qquad \text{(Eq 35)}$$

Where:

Prob(PE) = failure probability of electronic logic solver, assumed to be constant at 5E-3.

λ_{TS1} = failure rate of temperature switch 1, failures per unit time

λ_{TS2} = failure rate of temperature switch 2, failures per unit time

λ_{LS1} = failure rate of level switch 1, failures per unit time

λ_{LS2} = failure rate of level switch 2, failures per unit time

λ_{FT1} = failure rate of flow transmitter 1, failures per unit time

λ_{FT2} = failure rate of flow transmitter 2, failures per unit time

λ_{FT3} = failure rate of flow transmitter 3, failures per unit time

λ_{BV1} = failure rate of block valve 1, failures per unit time

λ_{BV2} = failure rate of block valve 2, failures per unit time

λ_{SOL1} = failure rate of solenoid valve 1, failures per unit time

λ_{SOL2} = failure rate of solenoid valve 2, failures per unit time

λ_{PT1} = failure rate of pressure transmitter 1, failures per unit time

λ_{PT2} = failure rate of pressure transmitter 2, failures per unit time

TI = test interval, time

The electronic logic solver PE is assumed to have a constant failure probability of 5E-3.

Failure rate data for the other devices are present in Table 4. The mode (most likely value) of the failure rate is set equal to the point value taken from the ISA example [Ref. 9]. The triangular probability distribution (Appendix A) was assumed to represent the failure rate data for the equipment items. The upper and lower limits are based on the failure rate data of Appendix 4 of Smith [Ref. 10].

Using the device failure rates represented by the Table 4 column labeled as the mode and the minimal cut sets presented in Table 3, the ISA technical report [Ref.9, page 35] computes an interlock PFD as 7.5E-3 (RRF of 133). This would satisfy the requirement for a SIL 2 interlock {PFD of 1E-2 or a RRF of 100}. Now what is the uncertainty in the interlock PFD?

The first thing to remember is that the PFD should be calculated based on the mean value of the failure rate lambda (λ), not the mode of the failure rate. For a data set, the mean is the best single point representation of the data set. Table 5 presents the PFD of each minimal cut set using the mean of the device failure rate lambda (λ). The total of the PFD becomes 1.47E-2 (RRF of 68). This would not satisfy a SIL 2 interlock requirement.

Now we will compute the variance of the predicted probability of top event, T. To compute the variance of the probability of the top event, T, we will need to compute the sensitivity of the top event to each of the variables.

Sensitivity Using Numerical Perturbation Calculation

We will use the ISA interlock example previously analyzed by Freeman (Ref. 7). Using public data sources Freeman found that the probability of failure on demand (PFD) was 1.49E-2. To use the perturbation method for the evaluation of the sensitivities requires evaluating the impact of small changes in the uncertain input parameters. To enable these calculations to be completed numbers using 8 significant digits were used. The calculations were completed using Microsoft Excel which uses double-precision floating-point arithmetic compliant with IEEE 754 specification. Excel nominally carries 15 significant figures in calculations. We will complete the calculations with Excel displaying 5 significant digits and round the result at the end of the calculations to 4 significant digits. We will illustrate the calculations by looking at the Flow Transmitter, FT1.

a. First we recalculate the PFD of the example and find it to be 1.4594E-02. This is the base number that we will use in the perturbation calculations.
b. We now make a small change in mean failure rate of FT1. The mean failure rate is 8E-6 failures/hour. A 1% perturbation is used or an increase of 0.08E-6 failures/hour.

c. The perturbed value of the failure rate of FT1 becomes (8 + 0.08) * 1E-6 failures/hour or 8.08E-6 failures/hour.
d. The perturbed value of the failure rate of FT1 is now used to re-calculate the PFD of the interlock. The resulting re-calculation finds that the PFD is 1.4618E-02
e. The change in the PFD from the base number due to the perturbation of the failure rate of FT1 is now computed as:

   delta PFD = 1.4618E-02 - 1.4594E-02 = 2.4556E-05

f. The sensitivity of the PFD to a change in the failure rate of FT1 is now computed by dividing the change in the PFD by the change FT1 failure rate as:

   sensitivity = (2.4556E-05)/(0.08E-6 failures/hour) = 307.0 hours
   Result is rounded to 4 significant digits.

We now note that in the paper by Freeman (Ref. 7) that equation 32 presents the sensitivity for FT1 as:

$$\frac{\partial Prob(T)}{\partial \lambda \text{FT1}} = 307.0 \text{ hr} \qquad \text{(Eq 32 of Ref 7)}$$

The answers are exactly the same. The sensitivity calculations for all of the devices are presented in Table 6. Note that in every case the sensitivity found using the analytical expression and the numerical estimate are the same. This is a general result which will be true for all basic event equipment failure rates found in a fault tree minimal cut set.

Once found, the sensitivities can now be used to compute the variance contribution of each particular device to the overall variance of the top event PFD in the same manner as Freeman previously presented.

The sensitivities of the top event probability are summarized in Table 6. We may now compute the variance contribution of each uncertain variable using Equations 1 – 4 in the methodology section of this paper. The relative variance contribution is expressed as a percent of the total variance of the top event probability. The top event variance is found to be 8.24E-6 (failures per hour)^2 or equivalently the top event standard deviation is 2.78E-3 failures per hour.

In this example the variance (measure of uncertainty) is dominated by the flow transmitters. The flow transmitters account for almost 2/3 of the uncertainty in the top event probability. Failures of the flow transmitters appear in 3 of the minimal cut sets. Since the uncertainty in the transmitter failure rates is large, the flow transmitters should contribute a significant amount to the variance of the fault tree top event.

We can use the variance in the top event to define the likelihood of achieving SIL 2 performance of the interlock. The calculations are done using the normal probability distribution. The normal distribution is tabulated as the standard normal distribution using a normalization factor Z.

The standard normal factor, Z, [Ref. 11] is defined as:

$$Z = \left[\frac{xi - E(x)}{\sigma}\right] \qquad \text{(Eq 36)}$$

Where:

σ = standard deviation. Note that the variance of a random variable is the square of the standard deviation of the random variable.

E(x) = Expected value of the random variable x_i

For the example, the expected value of the probability of the top event of concern was calculated as

$$E[Prob(T)] = 1.47E\text{-}2 \quad \text{(Eq 37)}$$

From Table 4, the standard deviation in the probability of the top event is

$$\sigma_T = 2.78E\text{-}3 \quad \text{(Eq 38)}$$

The target PFD for a SIL 2 interlock must not be worse than 1E-2 (RRF of 100).

We compute the normal distribution Z factor at the SIL 2 target value as

$$Z = (0.01 - 0.0147)/2.78E\text{-}3 \quad \text{(Eq 39)}$$

$$Z = -1.69065 \quad \text{(Eq 40)}$$

Tabulations of standard normal distribution are presented in most statistics books such as Meyer [Ref. 11]. For this value of Z, the corresponding probability that the interlock will achieve SIL 2 performance is 4.55 percent. We can also compute the risk reduction that the proposed interlock design will achieve.

What is the interlock probability of failure on demand (PFD) that we can be 95% certain that the interlock will provide? At the 95% level, the corresponding Z factor is 1.65 [Ref 11]. We now compute the corresponding PFD of the interlock that corresponds to this Z factor.

$$Z = \left[\frac{x95\% - E[Prob(T)]}{\sigma}\right] \quad \text{(Eq 41)}$$

In this case:

$x_{95\%}$ = the PFD that we can be confident that the interlock will achieve

$$\sigma_T = 2.78E\text{-}3 \quad \text{(Eq 42)}$$

$$E[Prob(T)] = 1.47E\text{-}2 \quad \text{(Eq 43)}$$

$$Z = 1.65 \quad \text{(Eq 44)}$$

Rearranging equation 40 to find $x_{95\%}$ we obtain:

$$x_{95\%} = E[Prob(T)] + Z\,\sigma_T \quad \text{(Eq 45)}$$

$$x_{95\%} = 1.47E\text{-}2 + (1.65)\,(2.78E\text{-}3) = 0.0193 \quad \text{(Eq 46)}$$

or a 95% certain risk reduction factor of

$$RRF = 1/PFD = 1/0.0193 = 52 \quad \text{(Eq 47)}$$

By the same logic there is 5% chance that the risk reduction is a PFD of 0.01 or an RRF of 99. In summary, this interlock is likely to perform as a mid-range SIL 1 interlock, not the SIL 2 interlock desired by process hazards analysis team.

5. Methodology for Fault Tree Uncertainty Analysis

We can now generalize the methods used in the above example to a proposed method for the evaluation of the uncertainty in a fault tree. The following step-by-step procedure is modified from that that previously published by Freeman (Ref 7) to indicate the use of numerical methods to compute the sensitivity. Step 8 shown in *italics* indicates the modification..

1. Create the fault tree using standard methods outline in the CPQRA book [Ref 1].
2. Determine minimal cut sets by hand or using standard computer software such as SAPHIRE [Ref 12].
3. Define the needed failure rate data for each basic event.
4. Define those basic events that are considered to be uncertain.
5. For the basic events that are uncertain, define the probability distribution and associated parameters needed to numerically define the probability distribution. Appendix B of this paper presents the needed description of 4 commonly used probability distributions. In many cases the triangular distribution will be selected. The minimum, maximum and mode (most likely) parameters will be needed for the triangular distribution.
6. Compute the mean of each of the uncertain variable
7. Compute the probability or frequency of the fault tree top event of interest using the mean value for each uncertain variable.
8. *Compute the sensitivity of the top event probability or frequency using the numerical methods outlined in this paper.*
9. Compute the variance of each of the uncertain variables
10. Compute the variance contribution of for each of the uncertain variables to the top event probability or frequency using equation 6.
11. Compute the total variance of the top event of interest probability or frequency by summing all of the contributions determined in Step 10.
12. Compute the variance contribution percent of each uncertain variable by dividing the variable contribution (Step 10) by the total variance of the top event of interest (Step 11).
13. Define the level of risk the project is willing to take. What chance will the project management accept for potential failure of the interlock to achieve the desired risk reduction? In the above example, I have used a 5% risk of failure or a 95% certainty that the interlock will achieve a risk reduction factor of 52. In the example, there is a very low probability that the interlock will achieve a SIL 2 PFD of 0.01 or an RRF of 100.

Alternately you can report the 90% range for the top event of interest. For the example the 90% range starts at the 5% RRF of 99 and there is a 95% certainty that that an RRF of at least 52 will be achieved. The expected risk reduction for this interlock is an RRF of 68.

6. Conclusions

A general procedure has been presented to quantify the uncertainty in calculations of failure frequencies using the fault tree methodology. The uncertainty analysis procedure is based on the application of propagation of error and variance contribution analysis techniques to the minimal cut sets created during a fault tree study. The procedure is simple and can be incorporated into standard fault tree analysis programs such as SAPHIRE [Ref 8]. This procedure is based on perturbation calculation of the frequency of the top event from changes in the uncertain variables. Incorporation of a numerical perturbation method into standard fault tree software such as SAPHIRE would be easy to implement and would provide the means for evaluation of the uncertainty in the quantified fault tree results.

For the example interlock presented in this paper, the uncertainty in the device failure rates degrades the predicted interlock SIL from meeting SIL 2 requirements to becoming a mid range SIL 1 interlock. This is a general result. Uncertainty in design parameters will reduce the likelihood that the equipment will achieve desired results. The design engineer has two choices:

- Accept the conclusion that the interlock will not perform as SIL 2
- Re-design the interlock using more reliable components to achieve the SIL 2 target

The variance contribution analysis shown in Table 7 indicates that the major contributor to the uncertainty in the interlock performance is the Flow Transmitters. Flow transmitters certified to SIL-2 or SIL-3 performance with lower failure rates could improve the calculated PFD of the interlock.

7. References

1. Center for Chemical Process Safety , *Guidelines for Chemical Process Quantitative Risk Analysis – 2rd edition*, American Institute of Chemical Engineers, New York, 2000.
2. Frank P Lees, Loss Prevention in the Process Industries – 2rd Edition, Butterworth, ISBN 0-7506-1547-8, Oxford, UK, 1996
3. US Department of Labor, Occupational Safety and Health Administration, CFR 1910.119 - Process Safety Management of Highly Hazardous Chemicals, 1910.119(e)(2)(vi).
4. Edward B. Haugen, *Probabilistic Approaches to Design*, John Wiley, New York, 1968
5. Raymond. Freeman and Angela Summers, "Evaluation of Uncertainty in Safety Integrity Level Calculations," *Process Safety Progress,* Published online in Wiley Online Library (wileyonlinelibrary.com) DOI 10.1002/prs. 11805, 2016
6. Raymond. Freeman, "General Method for Uncertainty Evaluation of Safety Integrity Level Calculations – Part 2 Analytical Methods," *Process Safety Process*, Published online in Wiley Online Library (wileyonlinelibrary.com) DOI 10.1002/prs 111915, October 2017
7. Raymond. Freeman ,"Error Propagation and Uncertainty Analysis: Application to Fault Tree Analysis", Process Safety Progress, Published Wiley Online Library (wileyonlinelibrary.com/journal/prs) DOI: https://doi.org/10.1002/prs.12080, June 29, 2019

8. US Nuclear Regulatory Commission, "Systems Analysis Programs for Hands-on Integrated Reliability Evaluations (SAPHIRE) – Version 8," NUREG/CR-7039, Vol 3, June 2011.
9. ISA, "Safety Instrumented Functions (SIF) – Safety Integrity Level (SIL) Evaluation Techniques Part 3: Determining the SIL of a SIF via Fault Tree Analysis, ISA Technical Report ISA-TR-84.00.02-2002 – Part 3, Approved 17 June 2002.
10. David J. Smith, *Reliability Maintainability and Risk – 8th Edition*, ISBN 978-0-08-096902-2, Elsevier, London, 2011
11. Paul L. Meyer, Introductory Probability and Statistical Applications, John Wiley, New York, 1972

Table 1. Comparison of Analytical Sensitivity with Numerical Approximation with Perturbation of 10% in xi for the equation: $y = F(xi) = Axi^n + B$

Order of Equation n	Equation for Analytical Sensitivity $\frac{\partial y}{\partial x} = nAxi^{n-1}$	Sensitivity with xi = 1 and A=1	Numerical Sensitivity with $\Delta xi = 0.1\, xi$	Error %
1	$\frac{\partial y}{\partial x} = 1A(xi)^0$	1	1	0
2	$\frac{\partial y}{\partial x} = 2A(xi)^1$	2	2.1	5
3	$\frac{\partial y}{\partial x} = 3A(xi)^2$	3	3.31	-10%
4	$\frac{\partial y}{\partial x} = 4A(xi)^3$	4	4.64	-16%
5	$\frac{\partial y}{\partial x} = 5A(xi)^4$	5	6.11	-22%
6	$\frac{\partial y}{\partial x} = 6A(xi)^5$	6	7.72	-29%
7	$\frac{\partial y}{\partial x} = 7A(xi)^6$	7	9.49	-36%
8	$\frac{\partial y}{\partial x} = 8A(xi)^7$	8	11.44	-43%
9	$\frac{\partial y}{\partial x} = 9A(xi)^8$	9	13.58	-51%
10	$\frac{\partial y}{\partial x} = 10A(xi)^9$	10	15.94	-59%

Error = (Analytical- Numerical)/Analytical

Table 2. Comparison of Analytical Sensitivity with Numerical Approximation with Perturbation of 1% in xi for the equation: $y = F(xi) = Axi^n + B$

Order of Equation n	Equation for Analytical Sensitivity $\frac{\partial y}{\partial x} = nAxi^{n-1}$	Sensitivity with xi = 1 and A=1	Numerical Sensitivity with $\Delta xi = 0.1\, xi$	Error %
1	$\frac{\partial y}{\partial x} = 1A(xi)^0$	1	1	0
2	$\frac{\partial y}{\partial x} = 2A(xi)^1$	2	2.01	-1%
3	$\frac{\partial y}{\partial x} = 3A(xi)^2$	3	3.03	-1%
4	$\frac{\partial y}{\partial x} = 4A(xi)^3$	4	4.06	-2%
5	$\frac{\partial y}{\partial x} = 5A(xi)^4$	5	5.10	-2%
6	$\frac{\partial y}{\partial x} = 6A(xi)^5$	6	6.15	-3%
7	$\frac{\partial y}{\partial x} = 7A(xi)^6$	7	7.21	-3%
8	$\frac{\partial y}{\partial x} = 8A(xi)^7$	8	8.29	-4%
9	$\frac{\partial y}{\partial x} = 9A(xi)^8$	9	9.37	-4%
10	$\frac{\partial y}{\partial x} = 10A(xi)^9$	10	10.46	-5%

Error = (Analytical- Numerical)/Analytical

Table 3. Basic Events in Minimal Cut Sets For Example

Cut Set No	BE 1	BE 2
1	PE	-
2	TS1	TS2
3	LS1	LS2
4*	FT1	FT2
5*	FT2	FT3
6*	FT1	FT3
7*	BV1	BV2
8*	BV1	SOL1
9*	BV2	SOL2
10*	SOL1	SOL2
11	PT1	PT2

* Minimal cut Set with same basic event as another cut set

Table 4. Lamda (λ) Failure Rate Data With Uncertainty Limits

Device	Minimum Failures per 1E6 Hours	Mode Failures per 1E6 Hours	Maximum Failures per 1E6 Hours	Mean Failures per 1E6 Hours	Lamda Variance (Failures per 1E6 Hours)^2
Flow Transmitters FT1, FT2 and FT3	1.0	2.9	20.0	8.0	18.3
Pressure Transmitters PT1 and PT2	1.0	2.3	20.0	7.8	18.8
Temperature Switch TS1 and TS2	3.0	7.6	20.0	10.2	12.9
Level Switch LS1 and LS2	2.0	4.6	20.0	8.9	15.8
Block Valves BV1 and BV2	0.2	2.3	10.0	4.2	4.4
Solenoid Valves SOL1 and SOL2	1.0	2.3	8.0	3.8	2.3

Table 5. Probability of Minimal Cut Sets Based on Mean Failure Rates

Cut Set No	BE 1	BE 2	PFD based on mean λ
1	PE	-	5.00E-03
2	TS1	TS2	2.00E-03
3	LS1	LS2	1.52E-03
4	FT1	FT2	1.23E-03
5	FT2	FT3	1.23E-03
6	FT1	FT3	1.23E-03
7	BV1	BV2	3.38E-04
8	BV1	SOL1	3.38E-04
9	BV2	SOL2	3.38E-04
10	SOL1	SOL2	2.77E-04
11	PT1	PT2	1.17E-03

Table 6. Sensitivity of Example System PFD to Basic Event Failure Rates Based on a 1% Perturbation

Basic Event Label	Device Type	Mean Failures per 1E6 Hours	Failure Perturbation of 1% of Mean Failures per 1E6 Hours	Perturbed Failure Rate Failures per 1E6 Hours	Total System PFD Using Perturbed Failure Rate of Basic Event	Change in Total System PFD	Sensitivity of PFD to Change in Failure Rate Hours
FT1	Flow Transmitter	8	0.08	8.08	1.461848E-02	2.455603E-05	307.0
FT2	Flow Transmitter	8	0.08	8.08	1.461848E-02	2.455603E-05	307.0
FT3	Flow Transmitter	8	0.08	8.08	1.461848E-02	2.455603E-05	307.0
PT1	Pressure Transmitter	7.8	0.078	7.878	1.460560E-02	1.167179E-05	149.6
PT2	Pressure Transmitter	7.8	0.078	7.878	1.460560E-02	1.167179E-05	149.6
TS1	Temperature Switch	10.2	0.102	10.302	1.461389E-02	1.995945E-05	195.7
TS2	Temperature Switch	10.2	0.102	10.302	1.461389E-02	1.995945E-05	195.7
LS1	Level Switch	8.9	0.089	8.989	1.460912E-02	1.519596E-05	170.7
LS2	Level Switch	8.9	0.089	8.989	1.460912E-02	1.519596E-05	170.7
BV1	Block Valves	4.2	0.042	4.242	1.460037E-02	6.445958E-06	153.5
BV2	Block Valves	4.2	0.042	4.242	1.460037E-02	6.445958E-06	153.5
SOL1	Solenoid Valves	3.8	0.038	3.838	1.459976E-02	5.832058E-06	153.5
SOL2	Solenoid Valves	3.8	0.038	3.838	1.459976E-02	5.832058E-06	153.5

Table 7. Variance of Top Event Probability Due to Uncertain Device Failure Rates

Device No	Device Type	Label	Sensitivity of Top Event Probability to Device Failure Rate Hour^{-1}	Variance of Device Failure Rate (Failures per 1E6 Hour)^{2}	Variance Contribution to Top Event Variance	Contribution to Top Event Variance %
1	Logic Solver	PE	-	0	0.00E+00	0%
2	Temperature Switch	TS1	195.7	12.9	4.94E-07	6%
3	Temperature Switch	TS2	195.7	12.9	4.94E-07	6%
4	Level Switch	LS1	170.7	15.8	4.60E-07	6%
5	Level Switch	LS2	170.7	15.8	4.60E-07	6%
6	Flow Transmitter	FT1	307	18.3	1.72E-06	21%
7	Flow Transmitter	FT2	307	18.3	1.72E-06	21%
8	Flow Transmitter	FT3	307	18.3	1.72E-06	21%
9	Block Valve	BV1	153.5	4.4	1.04E-07	1%
10	Block Valve	BV2	153.5	4.4	1.04E-07	1%
11	Pressure Transmitter	PT1	149.6	18.8	4.21E-07	5%
12	Pressure Transmitter	PT2	149.6	18.8	4.21E-07	5%
13	Solenoid valve	SOL1	153.5	2.3	5.42E-08	1%
14	Solenoid valve	SOL2	153.5	2.3	5.42E-08	1%

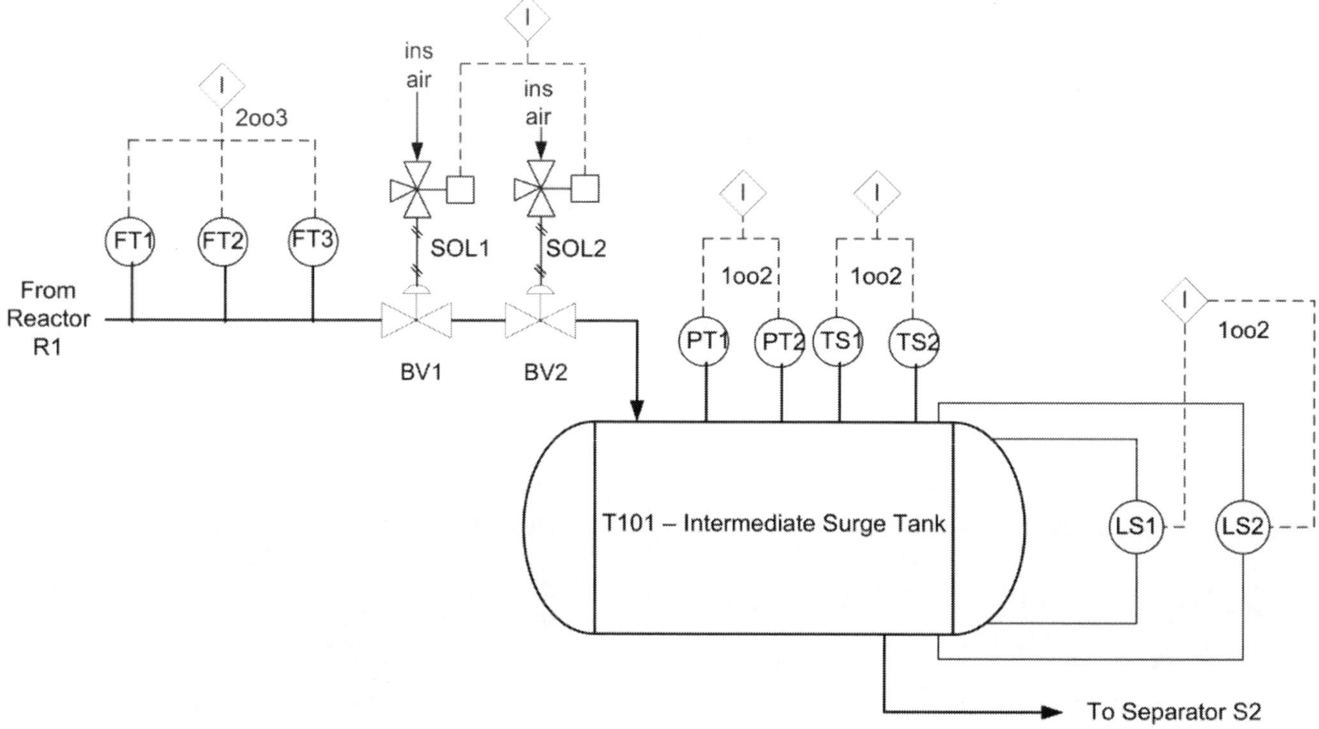

Figure 1. P&ID Diagram for Example

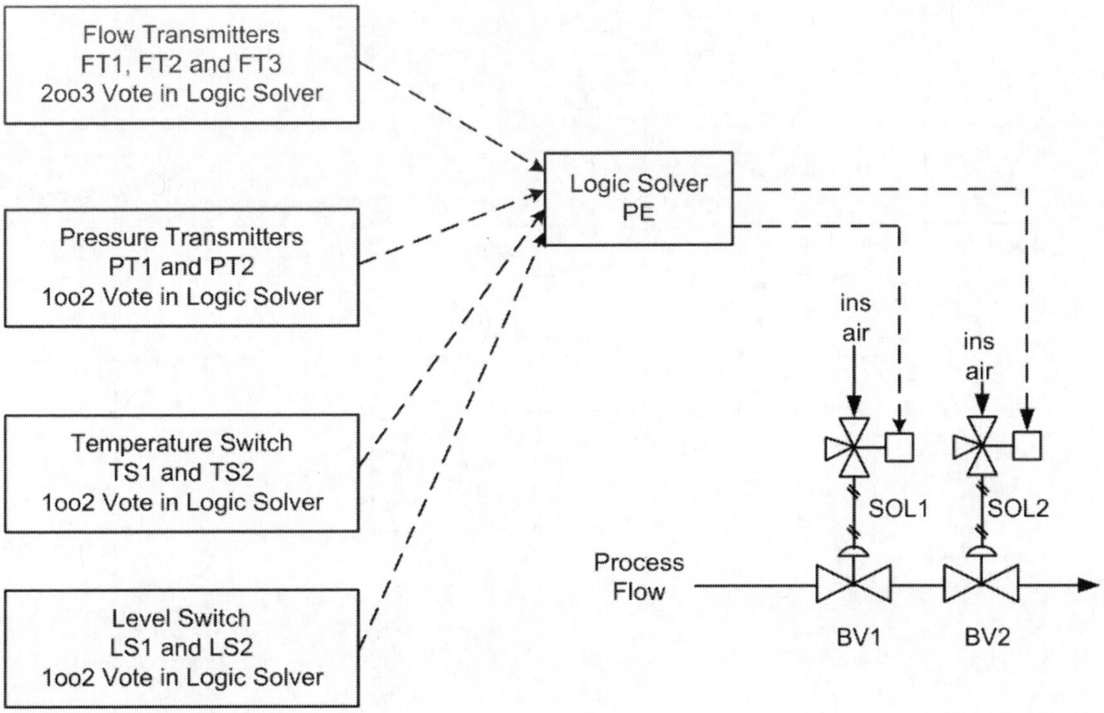

Figure 2. Interlock Block Diagram

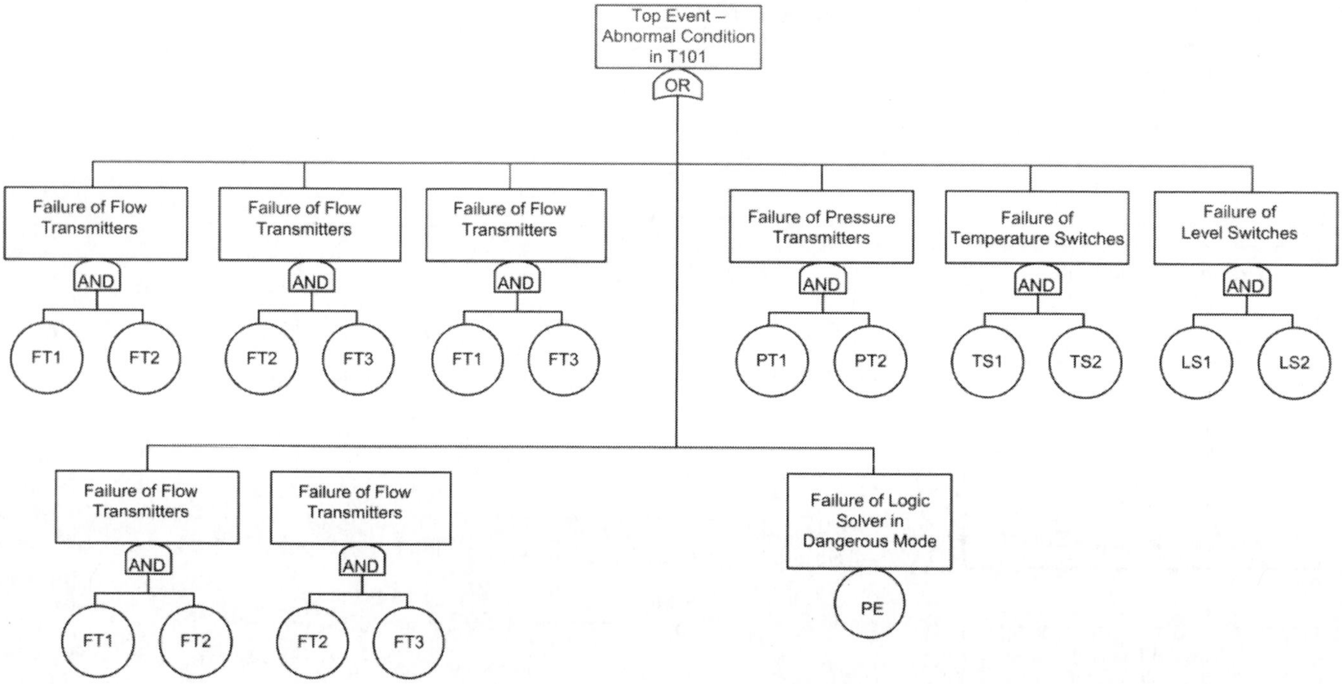

Figure 3. Fault Tree for Example Interlock

Appendix A – Commonly Encounter Probability Distributions

There are four probability distributions used to represent the failure rate of equipment items. These are:

1. Uniform Distribution
2. Triangular Distribution
3. Normal Distribution
4. Log-Normal Distribution

This appendix reviews the properties of these four probability distributions.

1. Uniform Distribution

The uniform distribution [Ref. A-1, pg 74; A-2, pg 687-688] is used when a variable is only known within a defined range. Figure A-1 presents a plot of the probability density function (pdf) for the uniform distribution. The distribution is defined using two parameters, A and B.

Where:

A = minimum value of the random variable

B = maximum value of the random variable

The mean or expected value of a random variable, x, that follows the uniform distribution is:

Mean = Expected Value – E(x) = (A+B)/2 (Eq A-1)

The variance of a random variable, x, that follows the uniform distribution is:

Variance = V(x) = $(B-A)^2/12$ (Eq A-2)

The standard deviation of a random variable, x, that follows the uniform distribution is:

Standard Deviation = $(\text{Variance})^{1/2} = (B-A)/(12)^{1/2}$ (Eq A-3)

2. Triangular Distribution

The triangular distribution [Ref. A-2, pg 686-687] is used when a variable is known to lie within a defined range and a "best guess" or estimate can be made as to the most likely value to the variable. Figure A-2 presents a plot of the probability density function (pdf) for the triangular distribution. The distribution is defined using three parameters, A. B and C.

Where:

A = minimum value of the random variable

B = maximum value of the random variable

C = most likely (mode) of the random variable

The mean or expected value of a random variable, x, that follows the triangular distribution is:

Mean = Expected Value – E(x) = (A+B+C)/3 (Eq A-4)

The variance of a random variable, x, that follows the triangular distribution is:

Variance = $V(x) = [A^2 + B^2 + C^2 - AB - AC - BC]/18$ (Eq A-5)

The standard deviation of a random variable, x, that follows the triangular distribution is:

Standard Deviation = $(\text{Variance})^{1/2} = \{[A^2 + B^2 + C^2 - AB - AC - BC]/18\}^{1/2}$ (Eq A-6)

3. Normal distribution

The normal distribution [Ref A-2, pg 665-666] is used when a variable when data analysis finds the normal distribution is the best model to describe the spread in the measured variable. Figure A-3 presents a plot of the probability density function (pdf) for the normal distribution. The distribution is defined using two parameters, μ and σ.

Where:

μ = mean of the measured data for variable

σ = standard deviation of the measured data for variable

The expected value of a random variable, x, that follows the normal distribution is:

Expected value = mean = $E(x) = \mu$ (Eq A-7)

The variance of a random variable, x, that follows the normal distribution is:

Variance = square of standard deviation = $V(x) = \sigma^2$ (Eq A-8)

The standard deviation of a random variable, x, that follows the normal distribution is:

Standard deviation = σ (Eq A-9)

4. Lognormal Distribution

The lognormal distribution [Ref. A- 2, pg 658-659] is used when a variable when data analysis finds the lognormal distribution is the best model to describe the spread in the measured variable. Figure A-4 presents a plot of the probability density function (pdf) for the lognormal distribution. Many variables in nature are lognormally distributed (for example the height of people). The log normal distribution is used when the variable of interest has a known physical lower limit of zero. A variable, x, is lognormally distributed if ln(x) is normally distributed. The lognormal distribution is defined using two parameters, μ_y and σ_y.

Let the original data be defined by the variable x with mean and standard deviation as:

μ_x = mean of the measured data for variable x

σ_x = standard deviation of the measured data for variable x

Define a new variable y as:

$y = \text{Ln}(x)$ (Eq A-10)

Then variable x is said to be lognormally distributed if y is normally distributed. We can now write the mean and standard deviation of y (μ_y and σ_y.) [Ref. A-3, A-4 and A-5] as

$$\mu_y = Ln\left[\mu_x^2 /((\sigma_x^2 + \mu_x^2)^{1/2})\right] \qquad \text{(Eq A-11)}$$

$$\sigma_y = \left[Ln(\sigma_x^2 / \mu_x^2 + 1)\right]^{1/2} \qquad \text{(Eq A-12)}$$

Where:

Ln is the natural logarithm (base e) of the argument

Appendix A References:

A-1. P. L. Meyer, *Introductory Probability and Statistical Applications*, 2nd Edition, Addison-Wesley, Reading Mass, Library of Congress Catalog no. 75-104971, 1972, pp. 74

A-2. D. Vose, *Risk Analysis–A Quantitative Guide*, 3rd edition, Wiley, Chichester, West Sussex, England, ISBN 978-0-470-51284-5, 2008, pp. 686-687

A-3. "Lognormal mean and variance - MATLAB lognstat". www.mathworks.com. Retrieved 27 December 2018

A-4. M. Mood, F. A. Graybill, and D. C. Boes. *Introduction to the Theory of Statistics.* 3rd ed., New York: McGraw-Hill, ISBN 0-07-042864-6, 1974. pp. 117.

A-5. E.L. Crow and K. Shimizu, Lognormal Distributions – Theory and Applications, Marcel Dekker, New York, ISBN 0-8247-7803-0, 1988, pp. 9

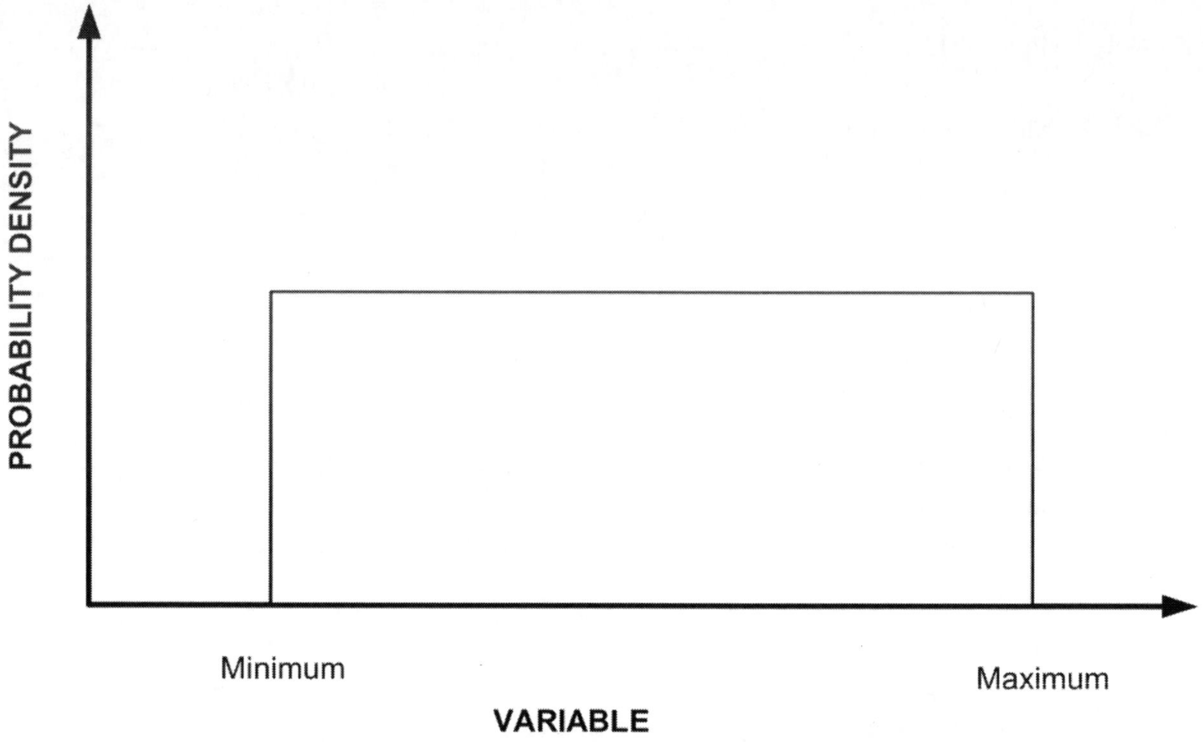

Figure A-1. Uniform Probability Distribution

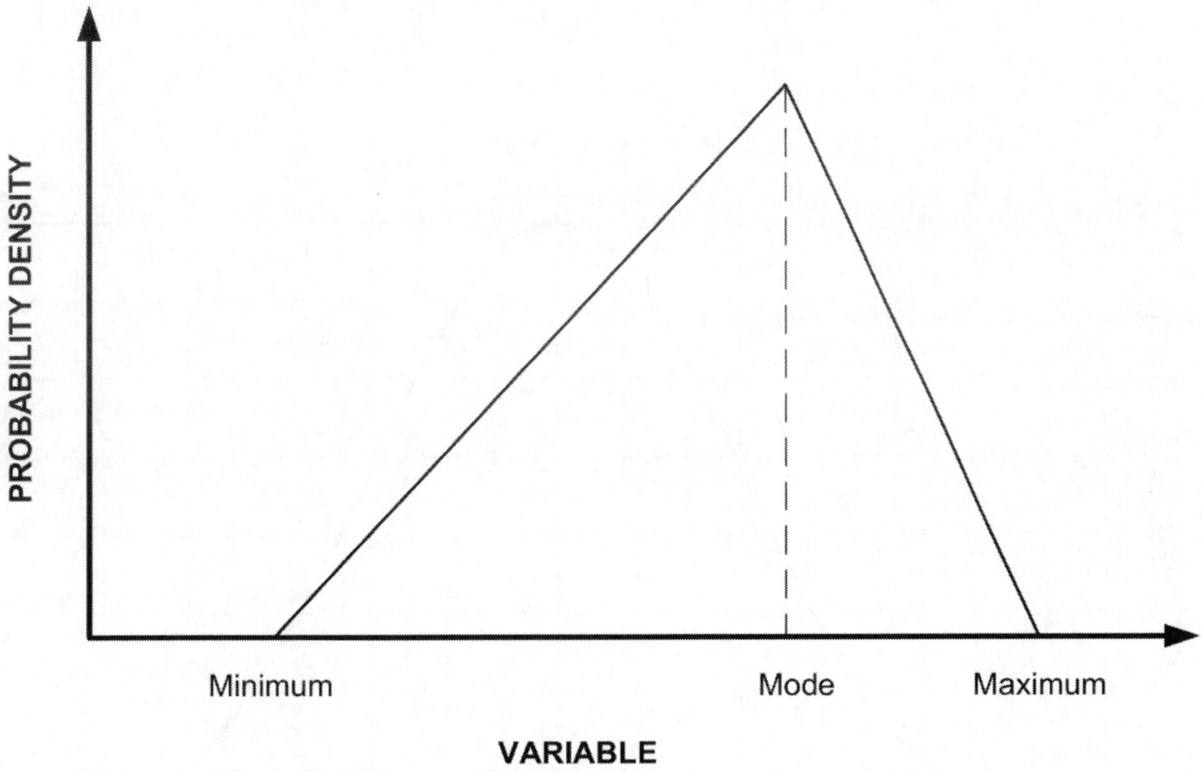

Figure A-2. Triangular Probability Distribution

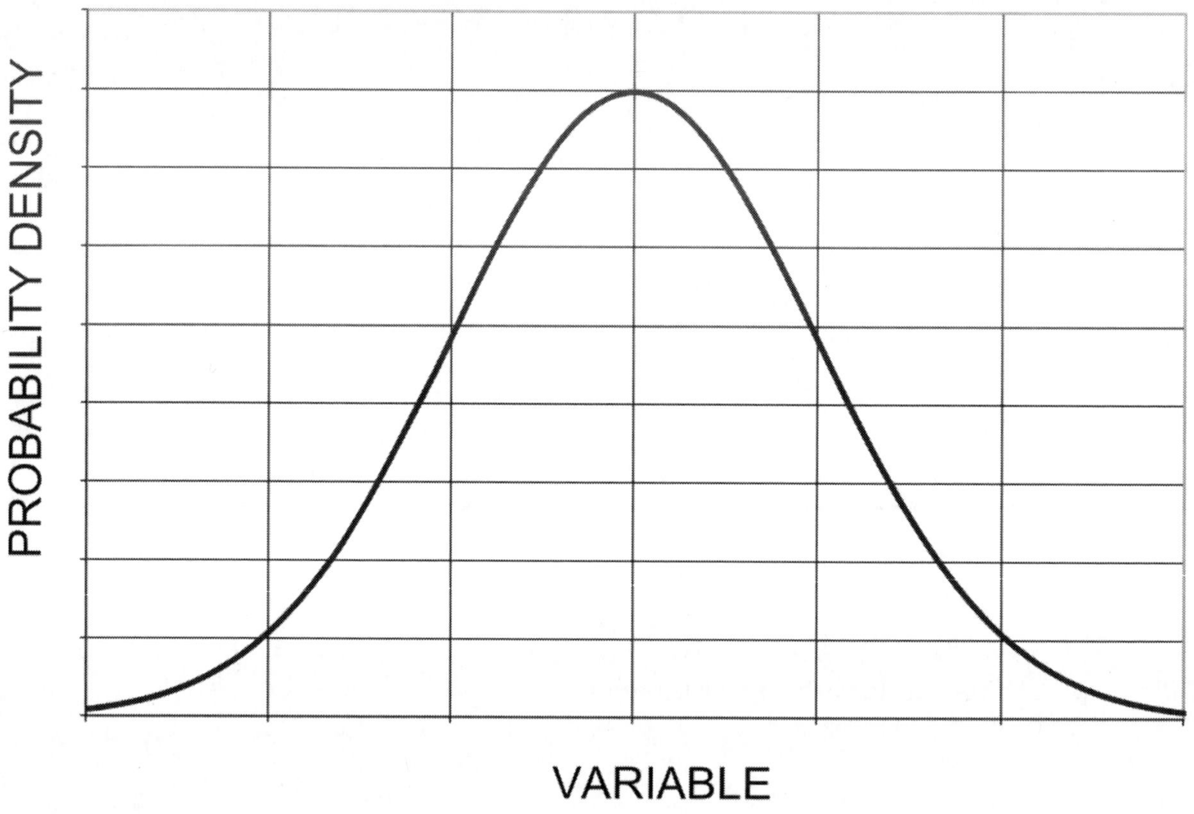

Figure A-3. Normal Probability Distribution

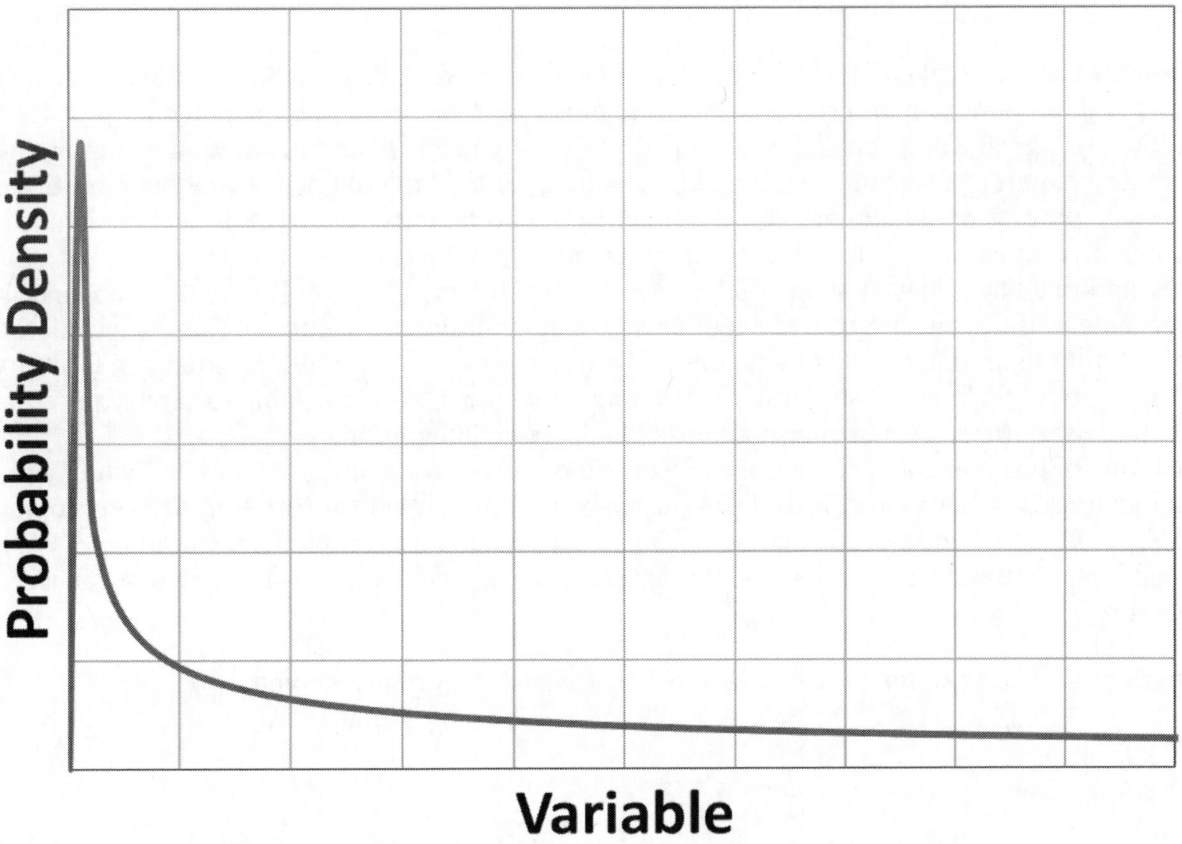

Figure A-4. Lognormal Probability Distribution

590424 Making the Most of Your PHA

Abstract:

A Process Hazard Analysis (PHA) is required to address OSHA Process Safety Management (PSM) and EPA Risk Management Program (RMP) regulations. Understanding the potential process hazard is widely recognized as a foundational element of PSM and RMP, and in addition to simply meeting regulatory requirements, can be a critical mechanism for flushing out potential vulnerabilities early in the design process for capital projects. Although an important activity, PHA is also recognized as a sizeable investment in time and energy for an Owner/Operator. So why not make the most of it? Other important activities at plants such as Mechanical Integrity, Alarm Rationalization, and Management of Change are able to stay focused and possibly be streamlined (i.e., reduced cost) based on the ability to use the results of a quality PHA having the necessary information and usability of documentation. So why not structure the PHA and documentation to be able to capitalize on potentially significant cost-savings and possible program integration?
The thrust of this paper and conference presentation will be to identify tips for conducting and documenting the PHA to maximize its ability to be used for supporting other important programs such as MOC, MI, and Alarm Rationalization. The objective will be to allow PHA Coordinators to be able to readily apply these tips to bring about improvements and cost-savings on these interrelated programs.

589814 Theoretical Analysis of the Viability of Isopropanol-Cyclohexane Solvent Systems for Biomatter Lipid Extraction

Abstract:

A theoretical alternative solvent combination for the extraction of lipids from biomatter is the Water, Isopropanol, and Cyclohexane instead of the more common combination used in the Bligh and Dyer Method (Water, Methanol, and Chloroform). While the Bligh and Dyer Method is considered effective, toxicological concerns have been raised over its use of both Chloroform and Methanol. The alternative method utilizes solvents that do not pose as high as a toxicological concern while theoretically maintaining a similar level of practical viability. The alternative method's extraction efficacy was judged against the performance of the Bligh and Dyer method using the simulation software Aspen Plus V10®. The equilibrium data generated from Aspen Plus V10® was then graphed into tertiary diagrams for data analysis using the graphing software SigmaPlot.

592166 Techno-Economic Analysis of the Conversion of CH4 and CO2 Intoacetic Acid Using Microwave Plasma

Abstract:

Methane and carbon dioxide together represent 90% of greenhouse gases [1], and one way of diminishing their emissions is through the development of pathways that convert them into added value chemicals. Acetic acid is one of such chemicals, and efforts at the laboratory scale have been carried out to obtain alternative ways of producing it. However, there are very few studies that evaluate the possibility of implementing these processes at a larger scale. In this paper, techno-economic and sustainability analyses were used to assess the feasibility of upscaling two pathways to produce acetic acid: (1) by direct conversion of methane and carbon dioxide and (2) by indirect conversion, using syngas (obtained from dry methane reforming) as an intermediate. For the direct method a microwave non-catalytic plasma reactor was considered as the main driver in the process simulation [2]. For the indirect method, a zeolite catalytical single step reaction unit was considered [3]. APEN Plus was used to simulate the reaction and purification systems of both methods, and alongside literature values and plant cost estimation techniques, the capital and operational expenses were calculated. Finally, the feasibility of both projects is assessed on carbon sequestration mass balances and economic metrics. The main sensitivity variables that need to be improved in order to obtain a minimum ROI threshold value were also evaluated.

Key words: carbon management, process design, microwave plasma reactor

[1] EPA. *Overview of Greenhouse Gases*. United States Environmental Protection Agency. 2017.

[2] Wang, L., Yi, Y., Wu, C., Guo, H., Tu, X. *One-Step Reforming of CO2 and CH4 into High-Value Liquid Chemicals and Fuels at Room Temperature by Plasma-Driven Catalysis*. Angewandte Chemie. 2017.

[3] Lopez, L., Velasco, J., Montes, V., Marinas, A., Cabrera, S., Boutonnet, M., Jaras, S. *Synthesis of Ethanol from Syngas over Rh/MCM-41 Catalyst: Effect of Water on Product Selectivity*. Catalysts. 2015.

589668 Model Predictive Control Design for the Allam Power Plant Integrated with an Air Separation Unit

Abstract:

Since electricity and heat generation account for around 25% of this greenhouse gas emission, high efficiency power generation technology such as the Allam cycle which emits essentially no CO_2 and NOx is much needed. Being an oxy-fuel power cycle, the Allam power plant is inherently integrated with an air separation plant (ASU). In this study Aspen Plus Dynamics was used to model the integrated Allam power plant/Air Separation Unit (ASU) with a high degree of heat and work integration based on the steady state work of Net Power. In our previous work, regulatory and advanced PID controls were implemented and tested for various set point tracking and disturbance rejection scenarios. However, PID controllers have limitations for various reasons. Consequently, a model predictive control (MPC) was designed in this study to optimize and improve the performance of the integrated unit and to test its performance in terms of dynamic behavior and robustness to model uncertainties.

589672 Explore the Function of Ion-Exchange Membrane (IEM) in Membrane Capacitive Deionization (mCDI) Cell, an Alternative Technique in Brackish Water Desalination Field

Abstract:

In the arid west, the freshwater supply of many communities is limited, leading to increased interest in tapping brackish water resources. Although reverse osmosis is the most common technology to upgrade saline waters, there is also interest in developing and improving alternative technologies. Here we focus on Capacitive Deionization (CDI), which has attracted broad attention as an energy efficient desalination technology, especially for ion extraction from low salinity brackish water. The objective of this work is to explore the function of ion-exchange membrane (IEM) in enhancing the performance of CDI cell. In this study, a fully coupled two-dimensional model describing ion transport and adsorption behavior in Membrane Capacitive Deionization (mCDI) cell was developed and solved numerically, taking advantage of Darcy's Law, Nernst-Planck equation, and Donnan equilibrium considering hydraulic dispersion in channel, non-ideal membrane, and modification of ion diffusivity in porous media. The model behavior was supported by the experimentally observed variations in transient effluent concentration under different operating conditions. The results showed that ion flux was significantly improved because of the IEM, leading to a faster desalination rate. The equilibrium adsorption amount is increased due to the permselectivity of the IEM, resulting in extra ion accumulation in electrode's macropores during desalination process. The cut-off time (optimal time to switch polarity to generate highest energy efficiency) of mCDI was shorter with a higher salt removal percentage compared to CDI under the same operating conditions. A sensitivity study of the thickness and the fixed charge density (starting from 500mM) of the IEM was done separately, and neither of them were found to be the dominant influencing factors under a low concentration range of brackish water (below 100mM) in this study.

590103 An Integrated Process for the Production of Propylene Glycol from Shale Gas

Abstract:

Propylene glycol (PG) is produced on a large scale primarily for the production of environmentally safest and least toxic industrial and domestic products such as polyester resins, thermoset plastics, polyurethanes, food additives and antifreeze. In this research, a novel integrated plant is developed for the production of PG from shale gas. This integrated approach has the benefit of safer operating conditions since the intermediate propylene oxide, which is explosive, does not need to be stored and transported. Furthermore, there are potential economic benefits from integration. The overall plant is simulated in the Aspenplus environment, and a variety of process conditions are tested at steady-state to optimize the production of PG. Heat-integration tools are utilized for energy-saving and capital cost reduction opportunities. A comparative economic assessment based on existing plant information indicates that the use of process integration techniques has the potential to reduce costs significantly.

Keywords: Process Integration, Heat Integration, Propylene Glycol

592456 Monitoring Methods for Fault Detection and Diagnosis: Algorithms and Applications

Abstract:

Many industrial processes collect an abundance of data from different sensors. These sensors often measure a wide variety of physical properties in order to ensure that these parameters are adhering to expected values. This is essential to ensure plant and operator safety, increase economic benefits, and maintain product quality.

A number of algorithms have been developed in order to improve existing fault detection and diagnosis performance. These algorithms integrate a number of different data-driven driven tools and methods. Multiscale wavelet-based representation of data can be used in order to handle data that is autocorrelated, non-Gaussian, and noisy. Hypothesis testing methods such as the Generalized Likelihood Ratio (GLR) technique can be used in order to provide the best possible detection for a fixed false alarm rate. Moreover, certain model-based methods have been developed in order to monitor process drifts and degradations in the process model, even when a process is operating under control.

This work will discuss the different tools utilized in development of the algorithms, their features, and practical applications, including the benchmark Tennessee Eastman Process (TEP).

589902 Selective Separation of Industrial Relevant Metal Ions from High Tds Water Using Wafer-Enhanced Electrodeionization (WE-EDI)

Abstract:

There is a need for an efficient and economical process of ion-selective separation. In hydraulic fracturing, for instance, several million gallons of water are used in a site, most of which are disposed of by deep well injection taking the water from the ecosystem and causing earthquakes. However, due to the high amount of ions in wells at fracking sites, pressure driven processes for water recovery are impossible to use. This project will demonstrate an emerging environmentally friendly Wafer Enhanced Electrodeionization (WE-EDI) technology which attempts to eliminate the shortcomings that come with normal EDI and separates effluent streams into concentrated ions and clean water. In WE-EDI, ion-exchange resins are replaced by wafers that consist of cation and anion exchange beads and a polymer binding agent. In this study, selective separation of these ion exchange wafers and the effect of electroosmotic drag on the membrane structure have been performed for various anions and cations (Na^+, K^+, Ca^{2+}, Cl^-) and improved selectivity and higher water recoveries have been observed while significant modeling has taken place to predict the selective separation as a strong function of the equilibrium constants between the ions and the beads. This project is supported by the Membrane Science and Technology (MAST) Center and is conducted with the collaboration and guidance of industry mentors.

593303 Influence of Particles on Amine Losses during CO_2 Capture: A Process Simulation Coupled Aerosol Dynamics Model _

Abstract:

Influence of Particles on Amine Losses during CO_2 Capture: A Process Simulation Coupled Aerosol Dynamics Model

David Dhanraj and Pratim Biswas

Aerosol and Air Quality Research Laboratory

Department of Energy, Environmental and Chemical Engineering

Washington University in St. Louis

St. Louis, MO 63130, USA

Efforts to reduce the concentration of atmospheric CO_2 have been initiated due to escalating concerns of global warming and climate change. Amine-based post-combustion CO_2 capture offers an economically viable retrofit for power-plants in combating these emissions. However, solvent losses have been a major challenge to commercializing this technology. Recent experiments[1] show that vapor-based and aerosol-based solvent losses are the key mechanisms of amine solvent loss during CO_2 capture. While vapor-based amine losses can be abated by employing techniques like water-washing, there has been little work to curtail aerosol-based solvent losses. This is primarily because the underlying mechanisms governing these losses are still not well understood. Although there are a few numerical models proposed for this system, there are no models that consider aerosol phenomena such as coagulation or particle loss. Moreover, these models have not been fully validated through experiments.

Furthermore, there are a few techniques that are being developed to tackle solvent losses. It is known that reducing particle number concentration in the flue gas will reduce amine emissions. Conventional Electrostatic Precipitators (ESP) are however inefficient in capturing particles in the size range 0.1 – 1.0 μm. Amongst different approaches demonstrated, introducing a pre-charging stage and a photo-ionization stage has shown potential in increasing the charging efficiency.

Therefore, the broad objectives of this work are primarily focused on two facets of the amine loss problem. The first is the development of a model that integrates aerosol phenomena with a process engineering model that accounts for the reactive mass transfer process in a packed-bed absorber. The second is the design, fabrication and testing of a bench-scale Photoionization Enhanced Two-staged Electrostatic Precipitator (PI-ESP) at Abbott Power Plant. In essence, with the PI-ESP aimed at reducing the particulate concentration in the flue gas, and the model relating the influence of particulate concentration in the flue gas to amine emissions, this work aims to illuminate the underlying mechanisms governing amine emissions, and demonstrate a solution to tackle the problem.

References

1. International Journal of Greenhouse Gas Control, vol. 85, pp. 86-99, 2019.
2. International Journal of Greenhouse Gas Control, vol. 19, pp. 138-144, 2013.

LSU College of Engineering Cain Department of Chemical Engineering

An Extension of Smoker's Equation That Includes the Murphree Tray Efficiency

2020 AIChE Spring Virtual Meeting
August 20, 2020
Paper 149a
11:00 – 11:30 AM

Brian Hanley
Dept. of Chemical Engineering
Louisiana State University
Baton Rouge, LA

LSU

LSU College of Engineering Cain Department of Chemical Engineering

Organization

- Are tray efficiencies still useful?
- Some reasons solutions to simplified problems are important
  - Misuse of vapor Murphree efficiencies in a process simulator
- McCabe-Thiele graphical procedure incorporating the Murphree vapor phase efficiency
- Smoker's equation for binary systems with constant relative volatility
- Adapting Smoker's equation to incorporate the Murphree vapor phase efficiency
- Examples
- Possible extensions

LSU

LSU College of Engineering Cain Department of Chemical Engineering

Are Tray Efficiencies Still Important? - 1

- What about rate-based simulations? Haven't they dispensed with the need for tray efficiencies?
  - Very few mass transfer correlations are available for trays.
  - Reliability of rate-based predictions over a range of systems, operating conditions, and tray types unknown.
  - Tray geometry choices are often not included or are included in very general ways.
  - Rate-based calculations are slow.
  - Rate-based calculations are often extremely sensitive to initial guesses/convergence criteria.
  - Not much published on the reliability of rate-based predictions for trays.

LSU

LSU College of Engineering Cain Department of Chemical Engineering

Are Tray Efficiencies Still Important? - 2

- Tray efficiencies (including Murphree) are still useful because…
  - Tray efficiencies have been reported for decades across a range of systems, tray types, and operating conditions.
  - They are relatively easy to measure (component efficiencies are more problematic).
  - They are, in theory, relatively easy to incorporate into solution algorithms originally designed for equilibrium trays.
  - These codes are fast and relatively insensitive to guesses/convergence criteria.
  - Some simulators/simulations can only be done with equilibrium trays or with code modified to include tray efficiencies (dynamic simulations).
  - Some companies are working on real-time data reconciliation tools that include the regression of tray efficiencies "on the fly" in an attempt to best match historical column data and predict future column behavior.

LSU

LSU College of Engineering Cain Department of Chemical Engineering

Usual Assumptions for Using Vapor Phase Tray Efficiencies

- The exit liquid stream is assumed to be at its bubble point.
- The exit vapor stream is assumed to be at its dew point.
- Since the vapor and liquid streams are not in composition equilibrium, the assumption of saturation above implies that *the two exiting streams are at different temperatures*, except for the case of equilibrium trays.
- There exists a fictitious vapor composition exiting a stage that is in equilibrium with the exiting liquid.
  - Vapor compositions determined by a bubble point calculation on the exiting liquid.
  - The fictitious vapor will be at the same temperature as the liquid but possibly at a different temperature than the actual vapor phase.

LSU

LSU College of Engineering Cain Department of Chemical Engineering

Murphree Vapor Phase Efficiencies - Definition

The Murphree vapor-phase stage efficiency for component "i" is defined by

$$E_{i,j}^{MV} = \frac{y_{i,j} - y_{i,j+1}}{y_{i,j}^{*} - y_{i,j+1}} = \frac{y_{i,j} - y_{i,j+1}}{K_{i,j}x_{i,j} - y_{i,j+1}}$$

$y_{i,j}$: average vapor phase composition of component "i" exiting stage "j"
$y_{i,j+1}$: average vapor phase composition of component "i" exiting stage "j+1"
$y_{i,j}^{*}$: (fictitious) vapor phase composition of component "i" exiting stage "j" determined by a bubble point calculation on the liquid exiting the stage

LSU

College of Engineering
Cain Department of Chemical Engineering

McCabe-Thiele Graphical Construction

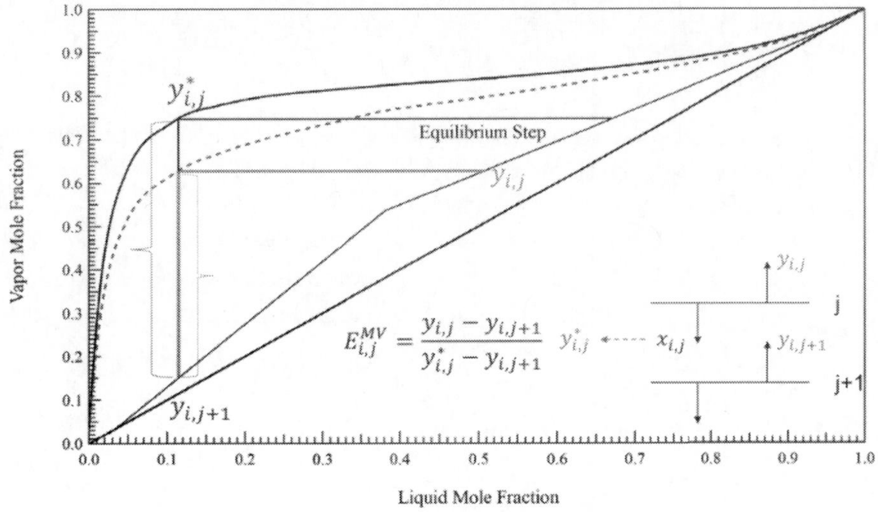

College of Engineering
Cain Department of Chemical Engineering

Binary Distillation: Smoker Equation for Finite Reflux Ratios

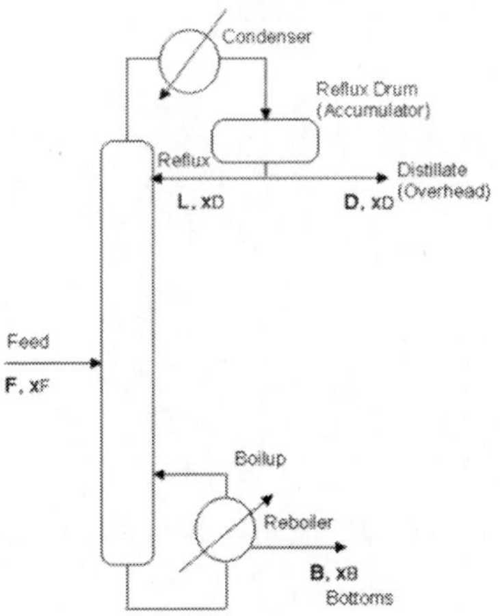

- Constant relative volatility, α

$$y = \frac{\alpha x}{1 + (\alpha - 1)x}$$

- Constant molal overflow

L constant
V constant

- Operating curves are straight lines: y = mx+b

$$m(\alpha - 1)x^2 + [\mathrm{m} + \mathrm{b}(\alpha - 1) - \alpha]x + b = 0$$

Call the root between 0 and 1 "k"

LSU College of Engineering Cain Department of Chemical Engineering

Binary Distillation: Smoker Equation for Finite Reflux Ratios

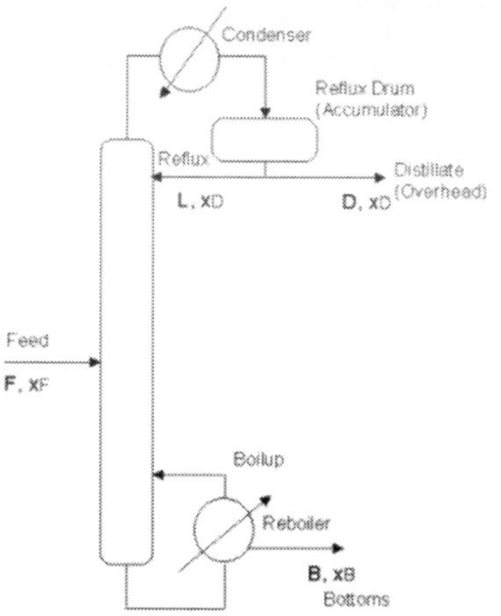

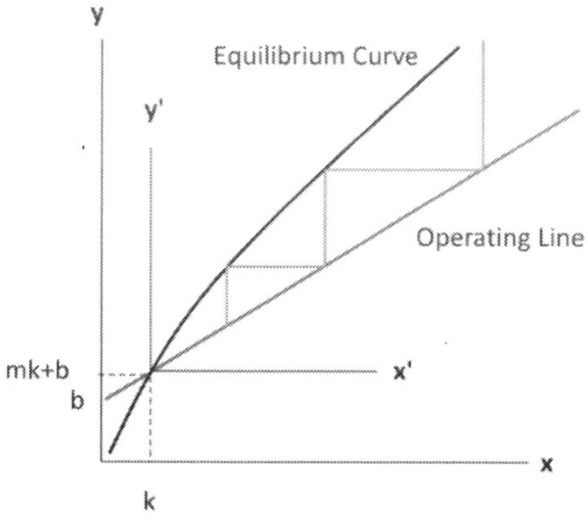

LSU

LSU College of Engineering Cain Department of Chemical Engineering

Smoker Equation for Finite Reflux Ratios – Rectifying Section Results

$$n_R = \frac{\ln\left\{\dfrac{x_0'\left[1-(mc(\alpha-1)x_n')/((\alpha-mc^2)\right]}{x_n'\left[1-(mc(\alpha-1)x_0')/((\alpha-mc^2)\right]}\right\}}{\ln\left(\alpha/_{mc^2}\right)}$$

$$m = \frac{R}{R+1}$$

$$b = \frac{x_D}{R+1}$$

$$x_0' = x_D - k$$

$x_n{}' = x_n - k$ ⟵ x_n = intersection of the operating line with the q-line

$m(\alpha-1)k^2 + [m + b(\alpha-1) - \alpha]k + b = 0 \qquad 0 < k < 1$

$c = 1 + (\alpha - 1)k$

LSU

LSU College of Engineering Cain Department of Chemical Engineering

Smoker Equation for Finite Reflux Ratios - Stripping Section Results

$$n_S = \frac{\ln\left\{\dfrac{x_0'\left[1-\left(mc(\alpha-1)x_n'\right)/\left((\alpha-mc^2)\right)\right]}{x_n'\left[1-\left(mc(\alpha-1)x_0'\right)/\left((\alpha-mc^2)\right)\right]}\right\}}{\ln\left({}^{\alpha}/_{mc^2}\right)}$$

$$m = \frac{Rz_F + qx_D - (R+q)x_B}{(R+1)z_F + (q-1)x_D - (R+q)x_B}$$

$$b = \frac{(z_F - x_D)x_B}{(R+1)z_F + (q-1)x_D - (R+q)x_B}$$

$$x_0' = x_0 - k$$ ← x_0 = intersection of the operating line with the q-line

$$x_n' = x_B - k$$

$$m(\alpha-1)k^2 + [m + b(\alpha-1) - \alpha]k + b = 0 \qquad 0 < k < 1$$

LSU

LSU College of Engineering Cain Department of Chemical Engineering

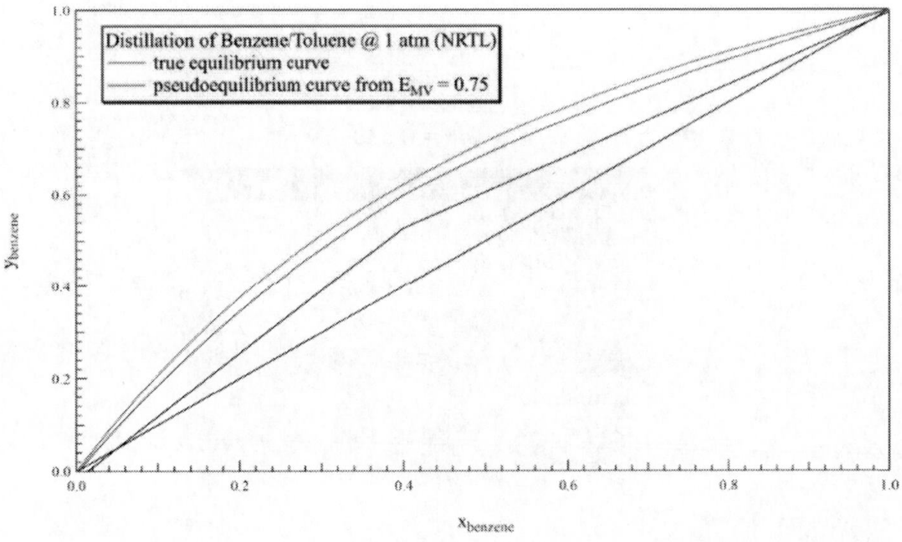

LSU

LSU College of Engineering Cain Department of Chemical Engineering

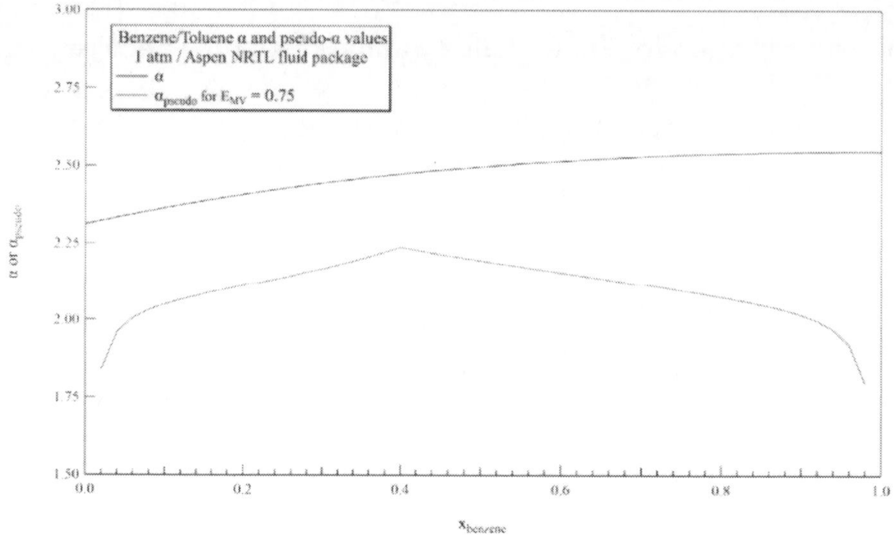

LSU

LSU College of Engineering Cain Department of Chemical Engineering

Benzene/Toluene Distillation @ 1 atm, α≈ 2.475

$z_F = 0.4$
$x_D = 0.975$
$x_B = 0.05$
RR = 3.5
D/F = 0.378

	McCabe-Thiele	Smoker
Total Stages	11	10.57
Feed Stage (from top)	7	6.01
	$E_{MV} = 0.75$	
Total Stages	15	14.69
Feed Stage (from top)	9	8.30

LSU

LSU College of Engineering
Cain Department of Chemical Engineering

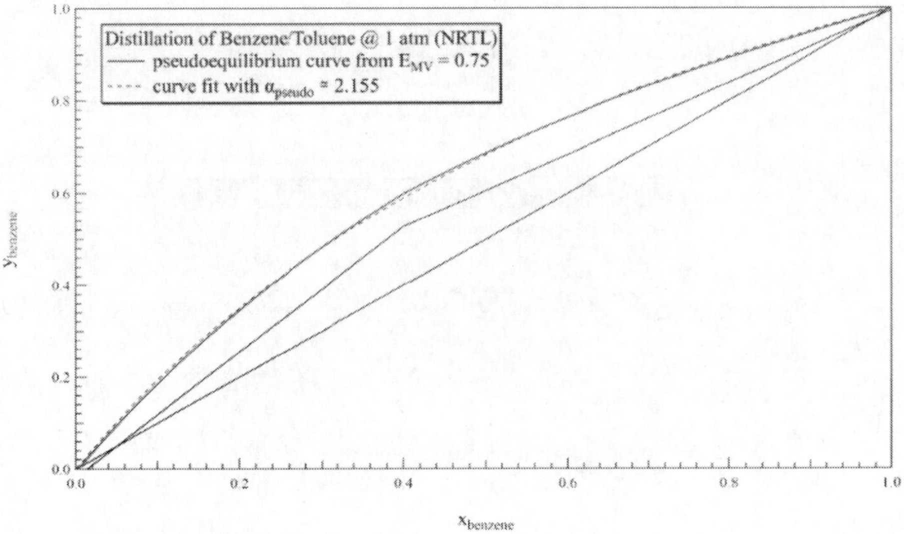

LSU

LSU College of Engineering
Cain Department of Chemical Engineering

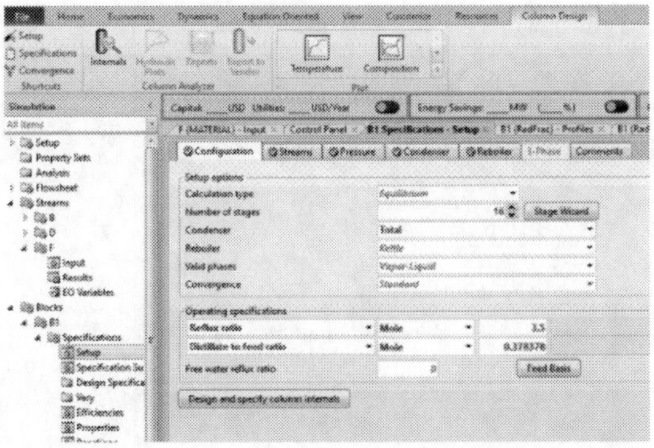

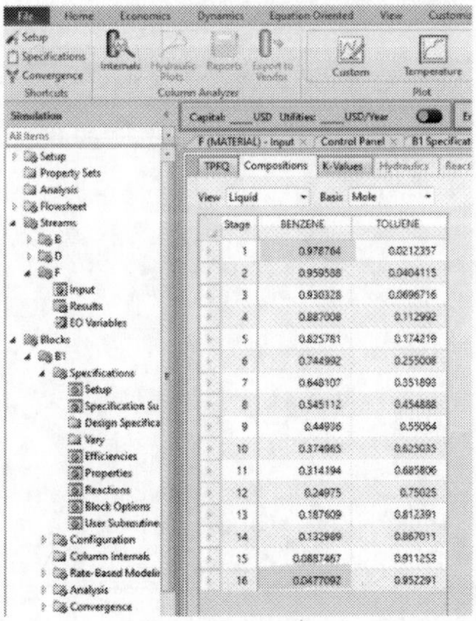

Stage	BENZENE	TOLUENE
1	0.978764	0.0212357
2	0.959588	0.0404115
3	0.930328	0.0696716
4	0.887008	0.112992
5	0.825781	0.174219
6	0.744992	0.255008
7	0.648107	0.351893
8	0.545112	0.454888
9	0.44936	0.55064
10	0.374965	0.625035
11	0.314194	0.685806
12	0.24975	0.75025
13	0.187609	0.812391
14	0.132989	0.867011
15	0.0887467	0.911253
16	0.0477092	0.952291

LSU

LSU College of Engineering Cain Department of Chemical Engineering

Argon/Oxygen Distillation @ 30 psia, α≈1.262

$z_F = 0.05$
$x_D = 0.999$
$x_B = 0.001$
RR = 130
D/F = 0.0491

	McCabe-Thiele	Smoker
Total Stages	?	81.3
Feed Stage (from top)	?	47
	$E_{MV} = 0.75$	
Total Stages	?	127.1
Feed Stage (from top)	?	64.23

LSU

LSU College of Engineering Cain Department of Chemical Engineering

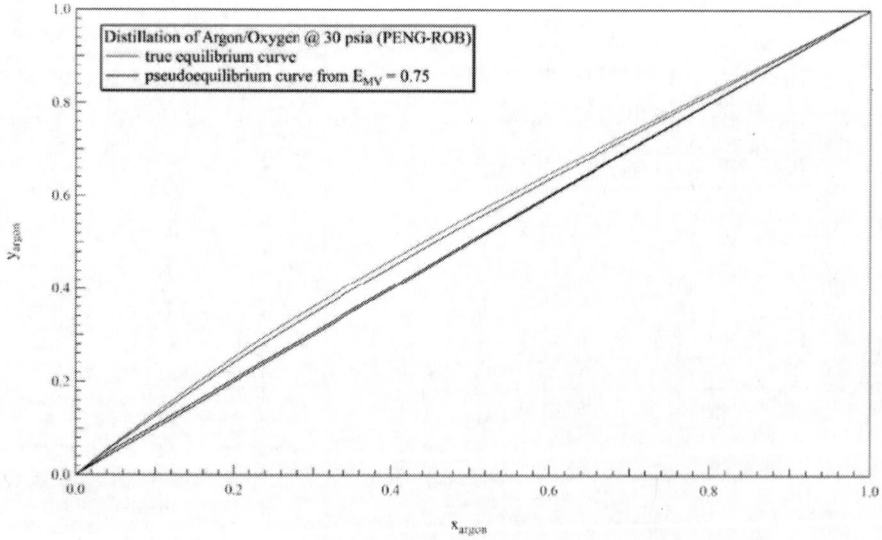

LSU

LSU College of Engineering
Cain Department of Chemical Engineering

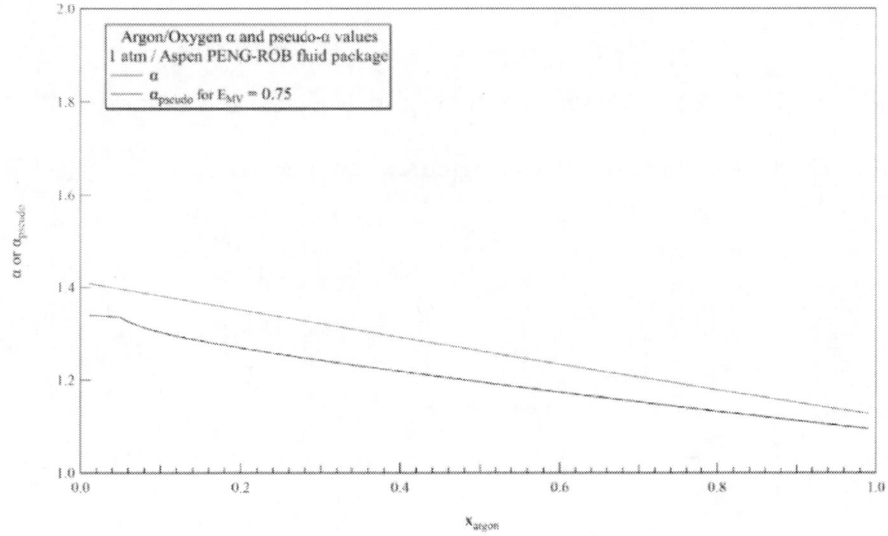

LSU

LSU College of Engineering
Cain Department of Chemical Engineering

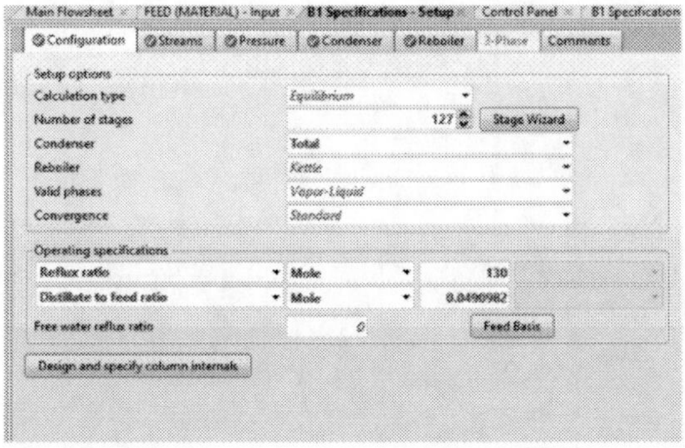

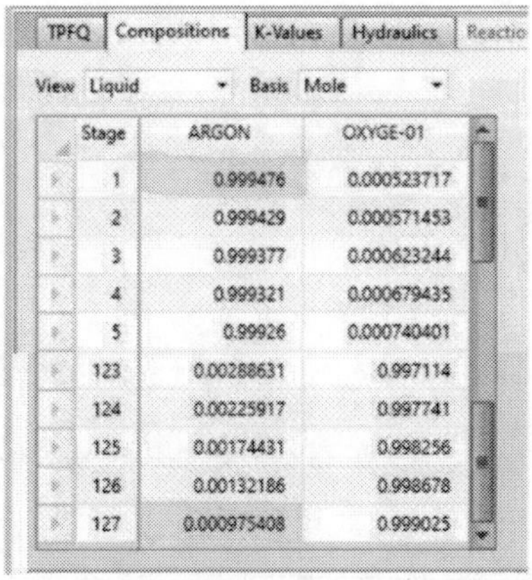

Stage	ARGON	OXYGE-01
1	0.999476	0.000523717
2	0.999429	0.000571453
3	0.999377	0.000623244
4	0.999321	0.000679435
5	0.99926	0.000740401
123	0.00288631	0.997114
124	0.00225917	0.997741
125	0.00174431	0.998256
126	0.00132186	0.998678
127	0.000975408	0.999025

LSU

LSU College of Engineering Cain Department of Chemical Engineering

Possible Extensions

- Fenske, Underwood, Gilliland method for multicomponent separations
- Analytical solutions for regressed values of the effective α's

LSU

LSU College of Engineering Cain Department of Chemical Engineering

- Thank you for attending
- I would be happy to answer any questions

Brian Hanley
Department of Chemical Engineering
Louisiana State University
3314C Patrick F. Taylor Hall
Baton Rouge, LA 70803

brianhanley@lsu.edu

LSU

LSU College of Engineering Cain Department of Chemical Engineering

Extra Slides on Misuse of Efficiencies in Process Simulators
(if time allows)

LSU

LSU College of Engineering Cain Department of Chemical Engineering

Why Even Simple or Approximate Models with Analytic Solutions are Important - Vaporization Efficiencies

The stage vaporization efficiency for component "i" is defined by

$$E_{i,j}^{V} = \frac{y_{i,j}}{y_{i,j}^{*}} = \frac{y_{i,j}}{K_{i,j} x_{i,j}}$$

$y_{i,j}$: average vapor phase composition of component "i" exiting stage "j"

$y_{i,j}^{*}$: (fictitious) vapor phase composition of component "i" exiting stage "j" determined by a bubble point calculation on the liquid exiting the stage

LSU

LSU College of Engineering Cain Department of Chemical Engineering

Analytic Solution for a Binary System at Total Reflux with Constant Relative Volatility Including the Vaporization Efficiency of the More Volatile Component

$$y_{1,n} = \frac{y_{1,R}(\alpha E_1^V - 1)(\alpha E_1^V)^n}{y_{1,R}(\alpha - 1)\left(\left[\alpha E_1^V\right]^n - 1\right) + \alpha E_1^V - 1}$$

$$x_{1,n} = \frac{y_{1,R}(\alpha E_1^V - 1)(\alpha E_1^V)^{n-1}}{y_{1,R}(\alpha - 1)\left(\left[\alpha E_1^V\right]^{n-1} - 1\right) + \alpha E_1^V - 1}$$

$y_{1,n}$: vapor phase mole fraction of the more volatile component exiting stage n
$y_{1,R}$: vapor phase mole fraction of the more volatile component exiting the reboiler
$x_{1,n}$: liquid phase mole fraction of the more volatile component exiting stage n
n: stage number (bottom up numbering)

LSU

LSU College of Engineering Cain Department of Chemical Engineering

How the "N[th]" Vaporization Efficiency Depends on the Other "(N-1)"

$$E_{N,j}^V = \frac{1 - \sum_{i=1}^{N-1} y_{i,j}}{1 - \sum_{i=1}^{N-1} \frac{y_{i,j}}{E_{i,j}^V}}$$

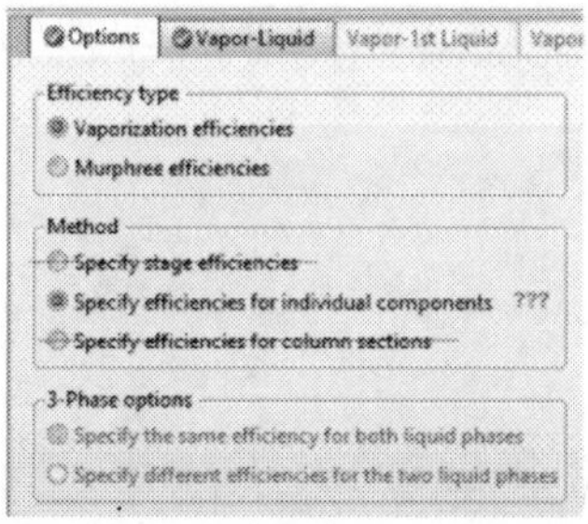

There are no limiting cases except $E^V_{i,j} = 1$ for every component

LSU

College of Engineering
Cain Department of Chemical Engineering

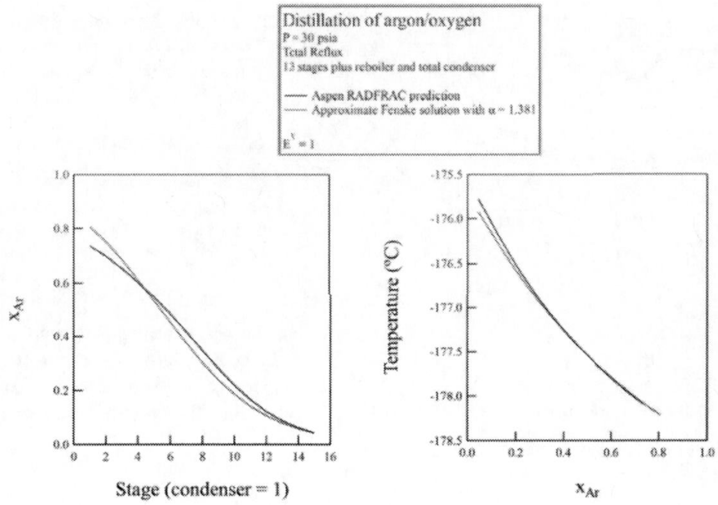

Aspen Plus v9

College of Engineering
Cain Department of Chemical Engineering

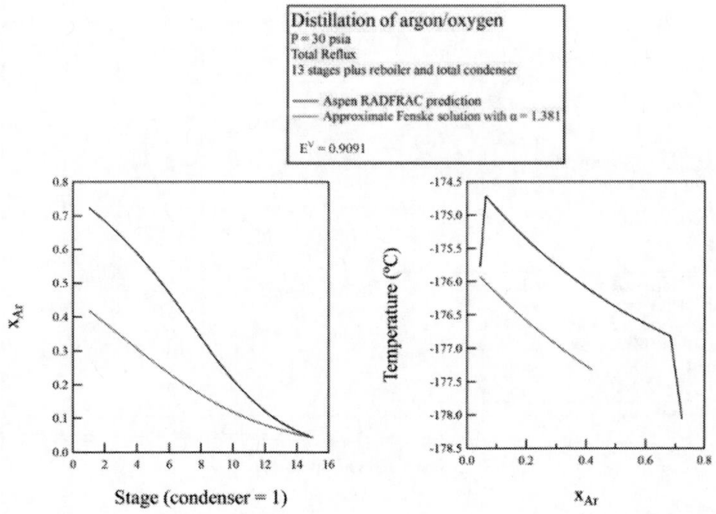

Aspen Plus v9

LSU College of Engineering Cain Department of Chemical Engineering

How the "N[th]" Murphree Efficiency Depends on the Other "(N-1)"

$$E_{N,j}^{MV} = \frac{\sum_{i=1}^{N-1} \left(y_{i,j} - y_{i,j+1} \right)}{\sum_{i=1}^{N-1} \frac{y_{i,j} - y_{i,j+1}}{E_{i,j}^{MV}}}$$

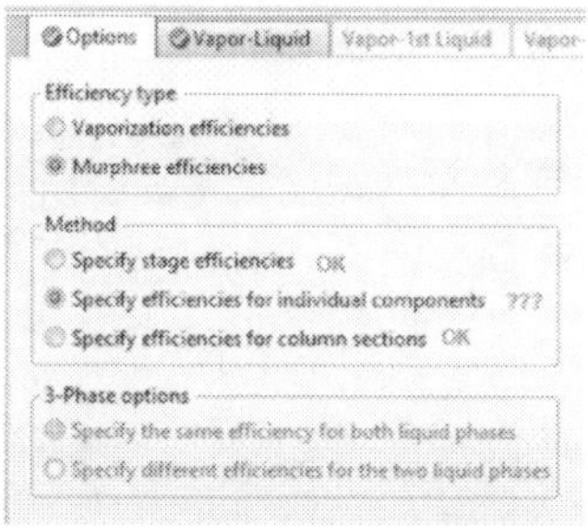

Limiting Cases:

Binary System

$E_{2,j}^{MV} = E_{1,j}^{MV}$

Multicomponent System with (N-1) E^{MV} Equal

$E_{N,j}^{MV} = E_{i,j}^{MV} = E_{j}^{MV}$

These limiting behaviors do NOT happen with the current Murphree component efficiency formulations in several commercial simulators

LSU

LSU College of Engineering Cain Department of Chemical Engineering

$$\sum_{i=1}^{N-1} y_{i,j}^{*} \neq 1$$

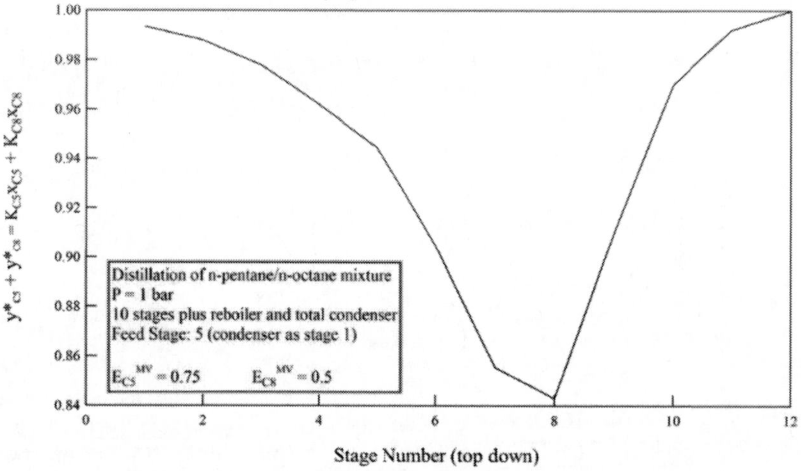

Aspen Plus v9

LSU

LSU College of Engineering
Cain Department of Chemical Engineering

HYSYS v9 Acid Gas Cleaning with MDEA Example†

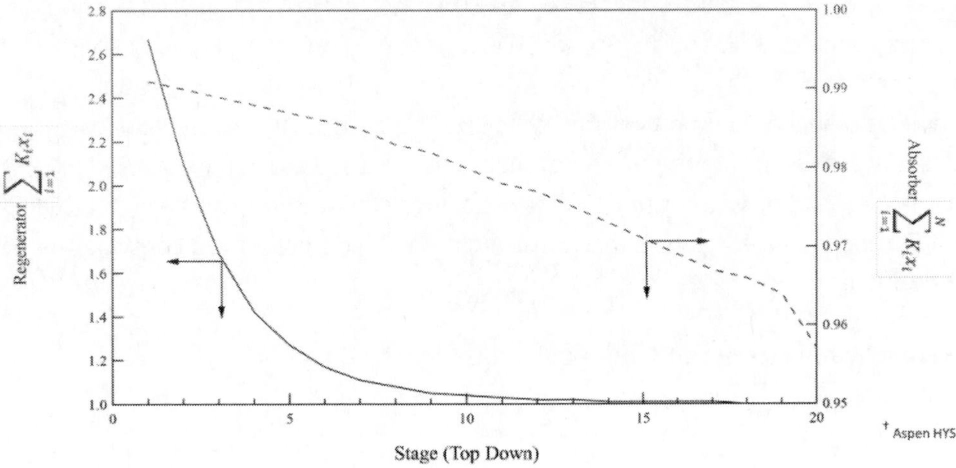

† Aspen HYSYS v9

LSU

590143 Debottlenecking of a Two Tower C2 Splitter

Abstract:

The importance of understanding both the capacity and efficiency of modern high performance trays in the successful revamp of distillation columns performing critical separations will be illustrated. A case study of the revamp of a two tower C2 Splitter located in an Ethylene plant operated by LG Chem in Yeosu, South Korea will be described in detail. This C2 Splitter was previously revamped twice using a combination of countercurrent and co-current trays. LG Chem Desired to maximize ethylene production and to reduce the slippage of ethylene being lost in the bottoms product. The case study will explain how the towers were subsequently revamped using a combination of high performance cross-flow SUPERFRAC® XT trays and co-current flow ULTRA-FRAC® trays at optimized tray spacings achieving an increase in ethylene production and an order of magnitude lower ethylene slip. Operating data will be presented showing the performance of the revised tower configurations.

589976 Distillation Column Performance Testing

Abstract:

Testing of distillation column internals, trays and packings, is still the best and reliable way of evaluating performances of distillation device. In this presentation, the role of the performance testing in separation engineering will be discussed. General approaches, procedures and guidelines for conducting, interpreting and reporting performance tests will be provided. The experimental techniques and principles of conducting the performance test will be presented in details.

590544 What Exactly Is Flooding Anyway?

Abstract:

What Exactly is Flooding Anyway?

Abstract

Flooding is commonly understood to determine the maximum capacity of a distillation column. However, most experts will be challenged if they are asked to provide a universal definition for flood. FRI has practiced the measurement and prediction of flood for various distillation scenarios for over 60 years. This study will debate the theory and phenomena encountered as a column approaches flood under different flow regimes. FRI's experience in control at or near operational flood will also be examined.

Key Words: Distillation, research, hydraulics, mass transfer, flood, operations

*Corresponding author. E-mail: Ken.McCarley@fri.org , Phone: 1-405-385-0354, Fax: 1-405-385-0357

588702 A Real-World Case Study: Distillation Tower Constraint Control

Abstract:

Distillation operations constitute a significant portion of energy usage and unit operations of every Hydrocarbon Processing Industry (HPI) facility. While the principles of thermodynamic and mechanical design of these unit operations are well understood, often times the design of the instrumentation and controls becomes the rate-limiting factor affecting their reliable performance. Further, as ultimately happens, economic drivers compel operators to maximize production from these units, thus requiring the controls to keep the units' operations within mechanical limitations, which further complicates the controls' design.

Historically, Model Predictive control (MPC) or "APC" was successfully implemented to ensure reliable and optimal operations at the mechanical limits of existing equipment. Sometimes, for a variety of reasons, however, MPC fails to generate the expected outcomes. A common observation made across multiple HPI sites and companies is that constraint-based regulatory controls are, poorly understood and, consequently, either not implemented at all or implemented with limited success.

This presentation uses a real-world example of a distillation tower that was subject to numerous operating constraints and illustrates the thought process behind designing the regulatory cont

589872 Simultaneous Design and Control of a Distillation Column Under Disturbance for Cost, Inherent Safety, and Controllability

Abstract:

Traditional inherent safety strategies in the early stages of design focus on reducing the overall hazard of a process without considering its operability. The process is first designed to be inherently safer with respect to a nominal, steady state case. Then, after the process is designed, layers of protection are added and operability issues are addressed.

However, the way a process is designed not only affects its safety but also heavily impacts its operability. An intensified process may contain less of a hazardous substance and thus be inherently less hazardous, but if the design restricts the controllability of the process, then the design may have a higher risk and be less safe overall. Therefore, in the early stages of design, it is necessary to consider of both the inherent hazard contained within the process and the ease by which these hazards can be controlled. With the dynamic behaviour of process systems becoming increasingly complex, the consideration of operability issues in the design stage becomes even more necessary to prevent incidents.

Simultaneous design and control is an effective method to predict and reduce operability issues that result from an uncontrollable process design. Incorporating traditional inherent safety metrics into simultaneous design and control subsequently allows for industry to investigate inherent safety in the early design stage along with an analysis of error tolerance and ease of control.

The objective of this research is to implement a strategy to simultaneously design and control an inherently safer distillation column. The PARametric Optimization and Control (PAROC) framework is used as a basis for the simultaneous design and control of a distillation column. gPROMS is used to perform dynamic optimization of a distillation column design. Multiparametric Model Predictive (mp-MPC) control is used to constrain the control limits of the control response, and the Safety Weighted Hazard Index (SWeHI) is incorporated into the PAROC framework to measure and constrain the safety of the system. Therefore, the distillation column is optimized for minimal cost under control and safety constraints.

589120 Multi-Objective Optimization of Dividing Wall Columns

Abstract:

Distillation is the most important separation technology in the chemical industry. In the USA it captures at least 2.5 % of the energy consumption [1,2]. Considering this significant energy expenditure and the environmental aspects, there is a high potential of improvement. Hence, research in the past years focused on reducing the energy demand of distillations. Intensified processes such as heat integrated distillation columns or reactive distillations are two examples for niche applications to save energy. Another more widely applied development of the past years are dividing wall columns [3]. This intensified column setup splits a ternary mixture into the pure components in just one shell by applying a partitioning wall in the tower. This results in up to 30 % less energy consumption compared to the direct split sequence assuming equal product specifications [4]. The actual savings are depending on many factors and cannot be calculated easily. In order to get a fair comparison to conventional setups, only optimal operational points should be compared.

However, the optimization of distillation columns is a multi-objective problem. This means there are typically several competing objective functions like the number of stages N, the reboiler duty Q and the product purities. Accordingly, the result of the optimization is not a single working point but a high number of non-dominated solutions called Pareto-optimal. Pareto-optimal points are characterized by the fact that they represent best compromises of the objective functions. If an objective function is to be improved by moving from one Pareto-optimal point to another, at least one other function deteriorates at the same time [5]. To obtain Pareto-optimal points, optimization variables influencing those have to be chosen and constraints are optional. In the case of dividing wall columns, the optimization variables are for example the feed and side draw stage, the liquid and vapor split above and below the dividing wall or the outlet streams.

A common way to optimize distillation setups is to minimize total annualized costs *TAC* [6,7]. The objective functions are combined to one cost function by using heuristic weighting factors and exponents. This limits the solution space significantly before the calculation and in most cases, only one optimal point is calculated being a very problem specific value. Like that, the results of a *TAC* minimization is barely applicable for another user since it represents only a very small part of the optimal solution space. To obtain a universally usable result, the whole solution space could be calculated first. Afterwards the user would be able to choose one of the possible, optimal operation points depending on the specific problem at hand. Like that, the calculation has to be performed only once and different operators can use the results several times.

Therefore, this works aims to achieve universal results of the optimization of distillation setups. The focus is on dividing wall columns and their comparison to conventional distillation setups.

For the optimization, an adaptive scalarization scheme is used [4]. In a first step it combines the multiple objective functions to one by using the weighted sum method. The single objective problem is then solved as for the *TAC* calculation, too. Unlike as for the *TAC* determination, the weights are changed after the calculation of an optimal point. This is done by an adaptive scalarization method called sandwiching. Like that, the minimum necessary number of optimal points is calculated to describe the Pareto-optimal solution space. For the process simulation data Aspen Plus® is coupled with the optimization tool.

After the optimization, the visualization of the multi-dimensional solution space is a big challenge. To overcome this problem, self-organizing patch plots are used [8]. In those plots, the exact solutions of the optimization are presented without any interpolation. Another advantage is that not only the objective functions can be shown but also the corresponding optimal variables. After the representation, filtering of the results is possible. Thus, the results can be adapted for each rectification setup of the same kind that is to be operated with the same feed and for the specific problems of the user. This method offers a much more all-purpose way of optimization.

In this contribution, the solution space of a dividing wall column will be presented in self-organizing patch plots. For that, only one optimization run had to be performed. With the help of different case studies, the universal applicability of the method will be presented.

- Literature

1. Humphrey JL, Keller GE. *Separation process technology*. Chemical engineering books. New York: McGraw Hill, 1997.
2. Energy Information Administration, National Energy Information Center. *Annual Energy Review 2001*. Washington DC, 2002.
3. Yildirim Ö, Kiss AA, Kenig EY. Dividing wall columns in chemical process industry: A review on current activities. *Separation and Purification Technology*. 2011;80(3):403–417.
4. Schultz MA, Stewart DG, Harris JM, Rosenblum SP, Shakur MS, O'Brien DE. Reduce costs with dividing-wall columns. *Chemical Engineering Progress*. 2002;98:64–71.
5. Bortz M, Burger J, Asprion N, Blagov S, Böttcher R, Nowak U, Scheithauer A, Welke R, Küfer K-H, Hasse H. Multi-criteria optimization in chemical process design and decision support by navigation on Pareto sets. *Computers & Chemical Engineering*. 2014;60:354–363.
6. Ge X, Yuan X, Ao C, Yu K-K. Simulation based approach to optimal design of dividing wall column using random search method. *Computers & Chemical Engineering*. 2014;68:38–46.
7. Ramírez-Corona N, Jiménez-Gutiérrez A, Castro-Agüero A, Rico-Ramírez V. Optimum design of Petlyuk and divided-wall distillation systems using a shortcut model. *Chemical Engineering Research and Design*. 2010;88(10):1405–1418.
8. Stöbener K, Klein P, Horsch M, Küfer K, Hasse H. Parametrization of two-center Lennard-Jones plus point-quadrupole force field models by multicriteria optimization. *Fluid Phase Equilibria*. 2016;411:33–42.

590678 High Purity Benzene Production from FCC Gasoline Using a Top Dividing Wall Column

Abstract:

High Purity Benzene Production from FCC gasoline Using a Top Dividing Wall Column

An increasing global trend among refiners to transition into refinery petrochemicals is attributed to a projected higher profitability of petrochemicals compared to refinery fuels. Aromatics are important building blocks in the petrochemical value chain, and most aromatics are derived from reformate or pygas. In addition, RFCC/FCC-based refiners have shown a disposition towards aromatics extraction from FCC gasoline. Benzene extraction, as opposed to saturation, not only gives value upgrade but also aids the refiner in MOGAS benzene content management.

As compared to reformate or pygas feedstocks, FCC gasoline has a higher quantity cyclic olefins and naphthenes. In traditional solvent-based extraction processes, these compounds are relatively difficult to separate from aromatics, imposing challenges on the extraction unit design. GT-BTX PluS® can circumvent this situation by employing a proprietary solvent blend with a strong aversion towards cyclic and normal olefins and high affinity towards aromatic species. The solvent's behavior, coupled with unique design features in the downstream hydro-desulfurization (HDS) unit, enables GT- BTX PluS® to achieve a benzene purity in excess of 99.8%. The benzene product obtained qualifies as petrochemical grade and requires no further processing such as re-extraction. A commercial unit with this technology, is operating in South Korea with the benzene product meeting all petrochemical grade requirements.

Leveraging on the success of the above-mentioned unit and operational knowledge gained, Sulzer GTC Technology has further improved the process by converting the stabilizer downstream of the HDS unit into a Top Dividing Wall Column (TDWC). Any aromatic saturation across the HDS unit coupled with slippage of cyclic compounds will directly affect the benzene purity. If not properly accounted for, this results in the requirement of a re-extraction unit to recover benzene purity. This presentation will show the benefits of using proper VLE along with a TDWC approach for azeotropic recycle back to upstream extraction section. This failsafe can be built into the unit to recover benzene purity to account for unwanted impurities in the benzene product.

590063 Lab-Scale Experimentation and Simulation of a Reactive Dividing Wall Column

Abstract:

Reactive dividing wall columns (RDWCs) are highly integrated systems that can simultaneously perform chemical reactions and multicomponent separations within the same vessel. While reactive distillation columns (RDCs) are well established and dividing wall columns are becoming increasingly popular in chemical processes due to their ability to present significant capital and operating costs savings, RDWCs have not been commercially adopted, in part due to the lack of experimental studies and validated models. Therefore, to increase public literature knowledge in this area, a complex aldol condensation reaction involving side reactions and two liquid phases is being studied at the lab scale. Discussion will focus on the lab-scale experimentation and validation of a RDC model for the system, use of results to properly deisgn a RDWC, and then initial results of the integrated RDWC process.

589987 Classification and Comparison of Dividing Wall Columns

Abstract:

Dividing wall columns (DWCs) are known to reduce capital cost of a distillation process by putting multiple columns into a single shell. However, in literature[1,2], DWCs are also said to reduce energy requirement, which is an incorrect statement since each DWC configuration requires same heat duty as its thermodynamically equivalent simple column configuration. For example, the well-known Wright's dividing wall column[3] is thermodynamically equivalent to the 3-component fully thermally coupled (FTC) configuration[4]. FTC configuration always has the lowest heat duty when all the columns are operated at similar pressures. Thus comparing Wright's dividing wall column with other three column configurations can often result in incorrect conclusion regarding heat duties.

Medenoor Ramapriya et al[4,5] presented a simple rule that, for the first time, enabled exhaustive enumeration of DWCs corresponding to any given thermally coupled distillation column-configuration in 2017. This work has enabled the synthesis of all possible DWCs for a given separation. However, use of "dividing wall column" continues to refer to only some common DWC configurations while ignoring several potentially useful DWC options.

In this research, all possible dividing walls are classified into 5 types, each type of dividing wall has its unique structural characteristics, such as the position of the dividing wall in the column, the number of breaking points on the column, the number of transfer streams, the number of condensers and reboilers, etc. These characteristics indicate the capital cost to build such a DWC. This classification is expected to simplify choices of dividing wall columns and provide a common language regarding DWCs.

A ternary separation is used as an example to illustrate how the classification works. The heat duty requirement of each ternary separation configuration has been calculated from the Global Minimization Algorithm[6] developed in the group. Combining energy consumption and capital cost, this research provides a guideline to choose the optimal DWC configuration for a ternary separation.

Literature Cited

1. Dividing-Wall Column for Fractionationof Natural Gas Liquids in Floating LiquefiedNatural Gas Plants. *Chemical Engineering Technology*. 2016;39(12):2348–2354.
2. Synthesis of dividing-wall columns (DWC) for multicomponent distillations—A systematic approach. *Chemical Engineering Research and Design*. 2011;89(8): 1281–1294.
3. Wright RO. Fractionation Apparatus. 1949. US Patent 2,471,134.
4. A Systematic Method to Synthesize All Dividing WallColumns forn-Component Separation—Part I. *AIChE J.* 2018;64(2):649-659.
5. Thermal Coupling Links to Liquid-Only TransferStreams: A Path for New Dividing Wall Columns. *AIChE J.* 2014;60(8):2949-2961.
6. Nallasivam, V.H. Shah, A.A. Shenvi, J. Huff, M. Tawarmalani, R. Agrawal, Global optimization of multicomponent distillation configurations: 2. Enumeration based global minimization algorithm, *AIChE J.* 62 (2016) 2071.

AUTHOR INDEX

AIChE
120 Wall Street, FL 23
New York, NY 10005-4020

ISBN 978-1-7138-2324-7

CURRAN ASSOCIATES INC.
proceedings.com